D1292311

CHILDREN OF THE DESERT

CHILDREN OF THE DESERT

THE WESTERN TRIBES OF CENTRAL AUSTRALIA

Volume One

GÉZA RÓHEIM

EDITED AND WITH AN INTRODUCTION BY
WERNER MUENSTERBERGER

BASIC BOOKS, INC., PUBLISHERS • NEW YORK

Library of Congress Catalog Card Number: 73–91079
SBN: 465–01042–3

Manufactured in the United States of America
74 75 76 77 78 10 9 8 7 6 5 4 3 2 1

I acknowledge with thanks the valuable assistance of Dr. Nicholas Peterson in the preparation of the discussion regarding the kinship structure. I also wish to express my warm appreciation to Dr. Derek Freeman for reading the manuscript.

W. M.

CONTENTS

INTRODUCTION

Other Realities: Psychoanalytic Research Among Primitive Peoples

BY

WERNER MUENSTERBERGER

A few days before his death (in 1954), Géza Róheim handed me a voluminous bundle of material gathered in 1929 during his field research in Australia. It was his unpublished work on the Pitjentara or, more generally, the Aranda and Lurittja among whom he had spent approximately eight months. Róheim considered this work his magnum opus and, having a premonition of his death, asked me to prepare his writing for publication.

A first field report was published in 1932.[1] He subsequently elaborated on his findings in numerous papers and in two books, *The Riddle of the Sphinx*[2] and *The Eternal Ones of the Dream*,[3] both works reflecting his theoretical psychoanalytic exposition based on his field observations.

Ever since Róheim had read *Totem and Taboo*,[4] Freud's study of primitive man's psychic make-up, the desert people of

Central Australia had been a source of vexation for him. He had no reservations about Freud's ingenuous assumptions of cultural origins, which bore the mark of the evolutionary point of view of such nineteenth-century thinkers as Haeckel, Lamarck, and Atkinson.

Brought up in a liberal Budapest milieu, Róheim had a pronounced sympathy for the extraordinary teachings of the Viennese doctor, especially with regard to his novel explanations of the origins of cultural ethics and religious beliefs, which made extensive reference to the most primitive tribes known, the Australian aborigines.

It might be said that Róheim's fieldwork in Central Australia is perhaps his most essential and basic contribution. It was carried out among the same people about whom he had previously written, but from a theoretical and possibly ideological viewpoint, unlike his earlier more factual perspective.

Freud and his contemporary, Sir James Frazer, had used secondary sources, on the grounds of which they had developed their vistas and concepts about the early stages of *homo sapiens* and man's apparent progression from primitivity to complexity in his thinking and behavior. Freud, having made revolutionary discoveries by close clinical studies, permitted himself considerable license as to psychoanalytic endeavors in the realm of anthropologic-historical processes. He spoke of the "burden of culture,"[5] that is, of human society being burdened by unconscious guilt about rebellion against and murder of "the primal father." Human culture was interpreted as an involved and multifaceted elaboration of an archaic event, a primeval murderous deed which, though not demonstrable, led to remorse, self-control, and eventually an agreed-upon social order.

The predominant aim of such speculations was to find a plausible explanation for moral restraint, obsessional religious practices, self-discipline, and social harmony, those embargos from without and within facilitating the various ways people live together. In his attempt to explain the roots of neurosis—and also the foundations of character formation—Freud recognized the price paid by mankind—the repression of gratification of instinctual drives. There is a causative connection between the requirements of civilization and man's discontents: "Sublimation of instinct is an especially conspicuous feature of cultural development."[6] This holds true for civilization itself, as it is of significance both to the individual in relation to self and in relation to others. Yet there is a fundamental qualitative difference between categories of sociocultural phenomena (such as religious beliefs, etiquettes, class divisions) and individually developed defenses and superego injunctions called upon as a safeguard against those instinctual forces which would endanger self-preservation.

Freud's universal ideology tended to obscure the unavoidable cultural and individual differences. This generalization was a result of psychoanalysis being a therapy based on empirical data and conceptualized in hypotheses, representing a device for studying and interpreting the extent and effect of unconscious processes and intrapsychic structure. However, the question arises as to how far clinically derived hypotheses can be validated outside of the two-person relationship provided by the unique clinical setting. Freud was of the opinion that variations of human life style and character do not contradict an essentially similar grammar of man's inner nature; that social configurations and patterns are an externalized articulation of a culture-specific intrapsychic conflict. The subtitle of *Totem and Taboo*, "Some Points of Agreement between the Mental Lives of Savages and Neurotics," underlines a methodological invalidation with which Róheim took issue, for it does not seem tenable to compare group action with an individual's internal sufferings.

Here is a logical contradiction. One cannot compare the mental condition of a well-functioning group with the pathology of an individual whose existence is characterized by malfunction within his own social environment. Moreover, to a certain degree behavior is patterned according to collective expectation. Thus, conclusions drawn about individuals without intimate knowledge of "racial and national traditions . . . as well as the demands of the immediate social *milieu*"[7] are bound to reflect subjective experiences of the investigator. It has to be observed that the psychic whole can only be adequately understood in terms of the historical era and the social reality in which the ego is presented with a code, direction, and alternatives in which and through which it can test its coherence and identity.[8] That the expectations and restrictions of a given society leave their discernible mark on defenses, secondary processes, and sublimations goes without saying. Thus the psychoanalytic anthropologist must chart a new technique, for it is imperative to take the culture of the informant into consideration. Even though Freud paid attention to the cultural spectrum and was aware of a link between the patient's social history and his symptomatology, he did not heed the implications of the specific sociocultural impact. Needless to say, no analyst can remain unaware of environmental influences on ego and superego structures, and on the management of instinctual demands, thought organization, and reaction formations.[9]

Today we have moved an impressive distance from several propositions in *Totem and Taboo,* particularly those pertaining to the function of totemism as the initial instrument of human morality. But Freud, like Frazer, had never encountered a primitive. They had been accustomed to interpretations along evolutionary lines, and saw psychosocial development in an almost mechanical sense as a series of logical successive steps. It was still impossible to view man in culture in the complex inter-relatedness of internal and external forces. Conceived in this schematic way, totemism was not grasped in its basically metaphorical scope and utilitarian function-serving interests which otherwise might run counter to the equilibrium of disparate biological and sociological pulls. Totemism or its equivalents in kinship rules can be seen as one of the many strategems mankind has created in the service of the renunciation of instinctual aims.[10]

The impact of renunciation of instinctual aims on sociocultural development presented itself as a paramount question closely interwoven with psychoanalytic anthropology. To recognize to what extent one's concrete environment plays a consequential part, we have only to remind ourselves that psychoanalytic case histories, without exception, make overt or covert reference to cultural elements, often intimately bound up with the patient's complaint. For example, when Freud told *The History of an Infantile Neurosis*[11] he gave a "General Survey of the Patient's Environment and of the History of the Case." This might be considered a mere background description, but he does present his readers with a sketch of life among the upper bourgeoisie in prerevolutionary Russia as perceived and recalled by his patient, the well-known "Wolfman." Of course, a patient's perspective stems largely from his accessible memories colored by his pathology, by screened vision and idealizations—no doubt honest and candid, though rarely completely accurate.

The actuality of "a case" does not differ too much from the actuality of one's first encounter in the field, when even the simplest rules and manners are unknown and must be explored. I recall when I gave a helpful young man on the island of Siberut a small box of matches in exchange for fish he had caught for my first meal. It was an act entailing several mistakes of which I had been unaware: the gift separated him from everyone else; I had "forgotten" the symbolic implication of matches and fire; inadvertently I had aroused envy and suspicion, not being conversant with the indigenous customs regarding gifts, reciprocity, and traditional prerogatives. By means of this act I had bound some of his libido to me and brought out a transference response which had more than one effect: it endangered his position in the group while the group (predominantly his age group) could not risk his attachment outside the more or less stable libidinally bound community. Initially, theoretical concepts and generalizations tend to recede into the background upon one's first confrontation with the seemingly unsurveyable complexity of the organism of daily life in a native surrounding.

The scope of psychoanalytic anthropology is wider than that of either clinical psychoanalysis or cultural anthropology, and before we can apply our special knowledge to the larger edifice of the science of man we must grapple with the task of methodology.

One is tempted to say that we must first understand the rational function of an act or gesture before we arrive at comparison, analysis, and theory. Once Róheim had met the Australian aborigines in their own habitat, he focused on the details of their everyday affairs, their social, economic, religiointellectual life; in other words, he took the logical step by observing the concrete and tangible. This is the only way to get closer to the nature of the individuals we want to understand. The field researcher is dependent on the immediate relationship with the natives about whom he has hypothesized. But once the real people come alive, patterning the complexity and variability of the observed and related phenomena demands both a point of view and a method. While almost all of the old field reports were merely descriptive and often impressionistic or normative, the psychoanalytically educated ethnographer will focus on empathizing with his informants within their sociocultural setting, or, in Erik Erikson's words, "their concepts and their ideals in a coherent design for living."[12] This embraces their world view, and their experience of the accidental and the common, the rational and the irrational, for even the primary contact with one's environment is influenced by prototypical elements of the culture in question.[13] With such perspective in mind, one can hardly underestimate the multitude of interplaying factors which clearly defy the concept of an "average expectable environment" or "average expectable stimulations" such as Hartmann had proposed.[14]

In studying the relevant data, Róheim took the classical approach by trying to outline the social structure or, rather, the fabric of the standardized mode of intra- and intergenerational obligations and restrictions, but then did not limit his description to the strait jacket of observable data. He proceeded to explore the unconscious determinants of the kinship organization. From there he focused on other phenomena of the established system which illuminated the aborigines' way of guaranteeing their meager existence in the Australian bush. However, the essential ingredients of tangible reality do not explain the entire spectrum of the cultural and emotional life, let alone the uniqueness of private fantasy.

Empathy with solitary imagination and, by implication, with the sublimities of dreams, hopes, despair—in other words, that voyage from the mundane to the spiritual—is the clinician's concern. It is a methodological dilemma with which Róheim struggled, and every psychoanalyst since Róheim who has tried by various means to overcome this problem, which seems to transcend impartial scientific, even psychoanalytic self-searching acumen, has had something of the same plight. If practical considerations induced Róheim to start with an overview or cultural inventory, implicit in his conceptual model were certain psychoanalytic criteria: the libido theory, and such universal inevitabilities as the triangularity attendant upon the human infant's prolonged dependence (the Oedipus complex in psychoanalytic parlance).

This theoretical point of view does not simply marry psychoanalytic propositions and field observation; it also provides us with a direction in our work. Róheim took more or less a diagnostician's attitude once he was acquainted with the complicated social organization.

As a clinician he was aware of the phenomenon of transference, which he described as "the foundation-stone of every clinical analysis. By a transference we mean a displacement of the original infantile trends, or rather a partial displacement of these trends, from the father or the mother to the person of the analyst."[15] It is indeed this method of observation which allows us to obtain data which are not directly articulated and which can be applied to elicit a good deal of hidden information and components of the informant's mental functioning, giving us a clearer understanding of behavioral phenomenology and dynamic unconscious forces. Knowledge acquired in this way and carefully interpreted may bring forth a response that often leads to insight and yields additional data. My young houseboy on Siberut Island was the son of a widow who was suspected of being a witch. Apparently he felt victimized by the unequal way he was treated by fellow companions and the potential danger stemming from his deviant position. The matches he had obtained from me were not seen simply as a token of reciprocity, but as a potential tool for mastery and identification with me as a *tuan*, a man of importance (under the circumstances). This incident taught me how profoundly he was in need of compensation for the misfortune of his uncertain social position and some of the intolerance he had to face.

More recent psychoanalytic-anthropological work has focused on the social and maturational determinants, conscious and unconscious, of individual development. A frequent and seemingly simple method is the detailed observation of child training using the prototypical genetic model as the crucial influence of the immediate environment on the child's character during the critical pregenital phases, the pedagogical importance of which has been convincingly shown. Róheim's orientation was different, and younger anthropologists and even psychoanalysts have accused him of mistaking mere id-oriented examination of myths, dreams, and the fantasies expressed in children's games for research of personality development.

Much of the psychic sphere of people can be gleaned from the stories they have learned from their ancestors and which they share by retelling, the narration linking the past with the future. Thus, mythology is both collective and uniquely individual. Insofar as it is collective, there is an innate correspondence in the symbolic imagery which each individual creates for himself. The fundamental human disposition is grounded in the child's need for nurturance and protection. This primary information is both individual and group-specific: each infant enters the world under the same biological condition, which is part of the human plight. With respect to the experience of facing the world, the initial form and function—predominantly in a biological sense—seem to be alike for each human being. It appears that this similarity is reflected in many mythological key themes. But then the dynamics of the living world produce a vast number of variations, or rather individuations.

Somewhat disillusioned, Róheim asked toward the end of his life:

> Is there such a thing as a psychoanalytic anthropology? From a practical point of view psychoanalysis and anthropology are two different professions. A psychoanalyst sits in his office and receives patients who want to be cured. An anthropologist goes to some tribe in the desert or jungle and tries to understand those people—who certainly do not want to be cured, indeed, they could teach us a lesson in happiness. However, there has been a tremendous change since 1915 when I first wrote "psycho-analytic anthropology." . . . Anthropology has been transformed at least in America, and "Personality and Culture" is certainly a new science. . . . From my own point of view of course, what has been gained in breadth has been lost in depth. With few exceptions . . . we have something new that is in many respects revealing but it is definitely not psychoanalysis.[16]

Essentially Géza Róheim was a romantic. He kept on his desk the *tjurunga* (the symbolic genital emblem) given to him by one of his main informants.[17] He entrusted it to me, together with the present manuscript, as a symbol of recognition and reward. Symbols in dreams, fantasies, and gestures remained to him the royal road to the unconscious, as he had learned from Freud and his teacher Sandor Ferenczi. And so, telling us about the people in the Central Australian desert, his emphasis is on folk tales, religious concepts, rites, and ceremonies, those promptings of the unconscious which to him were *fons et origo* of being human.

We have made progress in the field of social and anthropological psychoanalysis, but we must not forget that pioneers such as Róheim opened the gate and taught us that there is more to human nature than the banal exigencies of everyday life, when we pursue our expeditions into the uncharted regions of the mind.

February 1973
State University of New York
Downstate Medical Center
Dept. of Psychiatry (Div. of Psychoanalytic Education)

NOTES

1. Géza Róheim, "Psycho-Analysis of Primitive Cultural Types," *International Journal of Psycho-Analysis* 13 (1932): 1–224.
2. Géza Róheim, *The Riddle of the Sphinx* (London: Hogarth, 1934).
3. Géza Róheim, *The Eternal Ones of the Dream* (New York: International Universities Press), 1945.
4. S. Freud, *Totem and Taboo, The Standard Edition of the Complete Psychological Works of Sigmund Freud,* ed. and trans. James Strachey, 23 vols. (London: Hogarth, 1953–1964), 13.
5. S. Freud, "The Future of an Illusion," *Standard Edition,* 21: 42.
6. S. Freud, "Civilization and Its Discontents," *Standard Edition,* 21: 97.
7. S. Freud, "An Outline of Psycho-Analysis," *Standard Edition,* 23: 146.
8. Cf. Erik H. Erikson, "Ego Development and Historical Change," in

The Psychoanalytic Study of the Child, ed. Ruth S. Eissler et al., 24 vols. (New York: International Universities Press, 1945–1969), 2 (1946): 359–396.

9. Cf. Werner Muensterberger, "Psyche and Environment," *Psychoanalytic Quarterly* 38 (1969), pp. 191–216.

10. Claude Lévi-Strauss, in his *Elementary Structures of Kinship* (Boston: Beacon Press, 1969), has taken issue with Freud's provoking thesis, and argues about "the failure of *Totem and Taboo*. . . . [Freud] ought to have seen that phenomena involving the most fundamental structure of the human mind could not have appeared once and for all. They are repeated in their entirety within each consciousness, and the relevant explanation falls within an order which transcends both historical successions and contemporary correlations. Ontogenesis does not reproduce phylogenesis, or the contrary" (pp. 490f.). It is interesting to note that Lévi-Strauss's argument of 1949 follows almost literally the position Róheim took from a thoroughly psychoanalytic point of view, first in *The Riddle of the Sphinx*, Chap. 3, "The Ontogenetic Interpretation of Culture," pp. 158–174, and subsequently in a more detailed form in his paper, "The Psychoanalytic Interpretation of Culture," *International Journal of Psycho-Analysis* 12 (1941), pp. 147–169. Even though sometimes influenced by certain psychoanalytic hypotheses, Lévi-Strauss confines himself to a structural perspective (now and then with echoes of a Nietzschean picture of the world) dissolving human existence into polarities of Nature and Culture, the Raw and the Cooked. For a conclusive interpretation of the human condition see also Edmund Leach, *Lévi-Strauss* (New York: Viking, 1970).

11. S. Freud, *The History of an Infantile Neurosis*, *Standard Edition*, 17 (1955).

12. Erik H. Erikson, *Childhood and Society* (New York: Norton, 1964), p. 159.

13. Cf. Muensterberger, "Psyche and Environment," 192f.

14. Heinz Hartmann, *Ego Psychology and the Problem of Adaptation* (New York: International Universities Press, 1964), pp. 35, 46, 51.

15. Géza Róheim, "Psycho-Analysis of Primitive Cultural Types," *International Journal of Psycho-Analysis*, 13 (1932), p. 8.

16. Géza Róheim, *The Gates of the Dream* (New York: International Universities Press, 1952), p. vii.

17. Róheim, *The Eternal Ones of the Dream*, pp. 86f.

CHILDREN OF THE DESERT

1

THE KINSHIP SYSTEM

The Origin and Workings of the Various Class Systems

A study of the kinship system and its functioning gives us a basis for understanding the psychology of the Central Australian people and their society and culture. The Australian tribes which we are to consider can be roughly divided into those with a two-class kinship system, those with a four-class system, and those with an eight-class system. In order to determine the system used by a given tribe, it is not enough merely to count the number of class names they employ, for a tribe which nominally has a two-class system may function much like one with an eight-class system.

It would appear that the two-class system was the earliest form. Today, we find that the class names are spreading from the four-class tribes (that is, the Matuntara and the Aranda) to their neighbors, the Pitjentara, Mularatara, and Jankurpatara, who have named endogamic classes. The class names make their first appearance as personal names. Then, owing to contact with four-class neighbors, intermarriage, and frequent discussions on marriageable and nonmarriageable women, the terms are adopted to cover marriage customs which have already come into existence. A class, as demonstrated by Radcliffe-Brown,[1] is merely the systematization or naming of existing marriage customs. It would not be surprising if a future investigator, visiting the scanty remains of the southern Aranda, should find that they have adopted the eight-class names of their northern neighbors.

That the evolution should be from a two-class to an eight-class system is in accord with our theoretical knowledge. The tendency to restrict the number of women available for marriage is the result of an incest phobia. Enlarging the number of classes and restricting the number of classes among which intermarriage is possible effectively reduces the number of women available for marriage. The class names themselves were probably invented in the area of the Aranda tribe, and disseminated by them to the neighboring tribes. The fact that the Lurittya use the class names with male and female suffixes indicates that the terms have been borrowed from another group and modified to meet their own needs.

Since, in determining whether a tribe has a two-, four-, or eight-class system, we cannot simply count the number of class names which the people employ, we must find some other criteria. One essential point is whether Ego, a member of an Australian community, marries his cross-cousin (his mother's brother's or father's sister's daughter), or whether the children of cross-cousins marry (i.e., Ego marries his mother's mother's brother's daughter's daughter). In the former case we are generally dealing with a two- or four-class system, in the latter case with an eight-class system.

In a true two-class system, the father and the son (if descent is through the male) are competitors for the same women, as Figure 1 indicates.

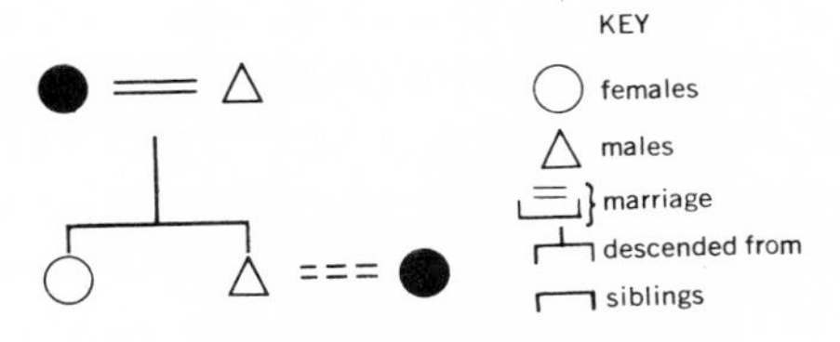

FIGURE 1

Cunow [2] has suggested that the four-class system arose from the two-class system by a combination of age grades and classes. In a four-class system the classes are so arranged that the father and the son belong to the same moiety, but to different classes. Figure 2 represents a four-class system with descent through the male line. Under this system, the father and the son are no longer competitors for the same women.

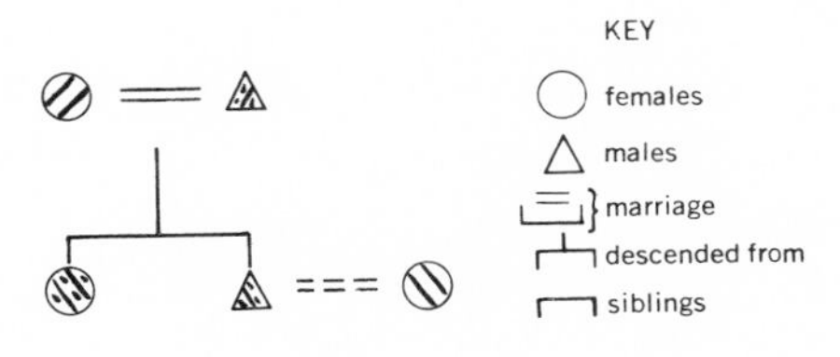

FIGURE 2

While both the father and the children are members of Moiety A, if the father is a member of Class I, the children are members of Class II. The original vertical division (that found in the two-class system) is exogamic; i.e., an *A* individual must marry a *B* individual. A new endogamic division splits the tribe into two endogamic classes (I and II). This is an age grade division. Parents and children are members of different classes, while grandparents and grandchildren are classed together.

The vertical, exogamic, division makes it impossible for brother and sister to marry, since they are members of the same moiety. The horizontal division has the same effect on parents and children. Whereas in some tribes (Dieri) only the vertical divisions are named, in others (Pitjentara) only the horizontal divisions are named.[3]

In practice, the differences between the Pitjentara four-class system and an eight-class system are not large. A comparison of Mularatara (a two- to four-class tribe), Pitjentara (four-class), and Matuntara (eight-class) terms shows the great similarities among the different systems.

English	Mularatara	Pitjentara	Matuntara
Father	mama	mama	katu
Mother	nguntju	nguntju	jaku
Elder brother	kuta	kuta	kuta
Younger brother	malantu	malantu	malantu
Elder sister	kankuru	kankuru	kankuru
Younger sister	malanytu	malanytu	malanytu
Son	katta	katta	katta
Daughter	untala	untala	untala
Aunt (father's sister)	kuntili	kuntili	kuntili
Uncle (mother's brother)	kamuru	kamuru	kamuru
Cross-cousin	watjira	watjira	watjira
Paternal grandfather	tamu	tamu	tamu
Maternal grandfather	kunarpi (tamu)	kunarpi (tamu)	kunarpi
Paternal grandmother	kami pakali	kami pakali	inkilji
Maternal grandmother	kami pakali	kami pakali	inkilji
Sister's son	okari	okari	okari
Sister's daughter	okari	no name	nankii
Mother-in-law (wife's parents)	waputu	waputu	waputu
Wife	kuri	kuri	kuri
Son's wife (husband's parents)	minkai	minkai	apili
Sister-in-law	toari	toari	toari
Brother-in-law	marutu	marutu	marutu
Mother-in-law (woman speaking)	kuntili	kuntili	ngumari
Mother's mother's brother	kami (no name)	kami (no name)	altali

In examining the relationships for which the terms differ, we must begin by omitting from consideration the words for mother, father, and son's wife. In these cases the differences are phonetic and have nothing to do with social organization. The three words cover the same degree of relationship—they differ only in sound.

Another group of dissimilar terms refers to members of the grandparents' generation. In the two-, the four-, and the eight-class systems, the same term is used to refer to the paternal grandfather. In none of these systems is this a relationship that is important to Ego. The same is not true of the maternal grandparents. Although the term used by the Matuntara for maternal grandfather (*kunarpi*) is known by the Mularatara and the Pitjentara, they rarely use it. They use instead the term for paternal grandfather (*tamu*). In the two- or four-class system the differentiation between maternal and paternal grandfather is not important. The Matuntara also differ from the other two groups in their term for maternal grandmother, and by having a special term for maternal grandmother's brother. This last is what one would expect in a group where the children of cross-cousins marry.

We can see from the following table the classes which exist in typical four- and eight-class groups, and also the groups among which intermarriage is permitted, as well as the class of the offspring of each union.

MEN		WOMEN		CHILDREN	
8-class	4-class	8-class	4-class	8-class	4-class
Purula Ngala	Purula	Pananka Ngaraii	Pananka	Kamara Mbitjinba	Kamara
Kamara Mbitjinba	Kamara	Paltara Pangata	Paltara	Purula Ngala	Purula
Pananka Ngaraii	Pananka	Purula Ngala	Purula	Pangata Paltara	Paltara
Paltara Pangata	Paltara	Kamara Mbitjinba	Kamara	Ngaraii Pananka	Pananka

A diagram given by Strehlow[4] shows the groups of relationship terms used among the various classes. The southern group of tribes (i.e., those with a four-class system) do not employ different terms for members of the Purula and Ngala, Kamara and Mbitjinba, Pananka and Ngaraii, and Pangata and Paltara classes. In actuality, when a Ngala migrates from the northern (eight-class) area to the southern (four-class) area (e.g., that of the Aranda tribe), he is classed by them as a Purula, by the Mbitjinba as a Kamara, and so on.

We can see from the table on p. 7 that the classes to the left of the brackets are co-classes from the southern Aranda point of view. The relationship terms by which the co-classes are distinguished from each other are missing in both the Aranda and Pitjentara systems. The Pitjentara do not differentiate in their relationship terms between the class from which the spouse is taken and its co-class. Although they have only two named classes, they do differentiate in their relationship terms among four groups of individuals, and are therefore a four-class tribe.

The Aranda, on the other hand, while they nominally possess a four-class system, in reality have an eight-class system, with four named and four unnamed classes. As a matter of fact, the southern Aranda Purula regards some Purulas as his brothers and others as his half brothers, so it would seem that the southern Aranda is aware that he calls what are really two classes by the same name. In the same way, he (as a Purula) regards some Pananka women as potential spouses (*noa*), while the rest (who correspond to the Ngaraii class in the north) he regards as his cross-cousins.

The genealogy in Figure 3 shows how an eight-class system functions. As can be seen, the class of any relative is known. Any individual belonging to the same class as the mother-in-law, for example, is considered a mother-in-law, and so on, through all possible relationships.

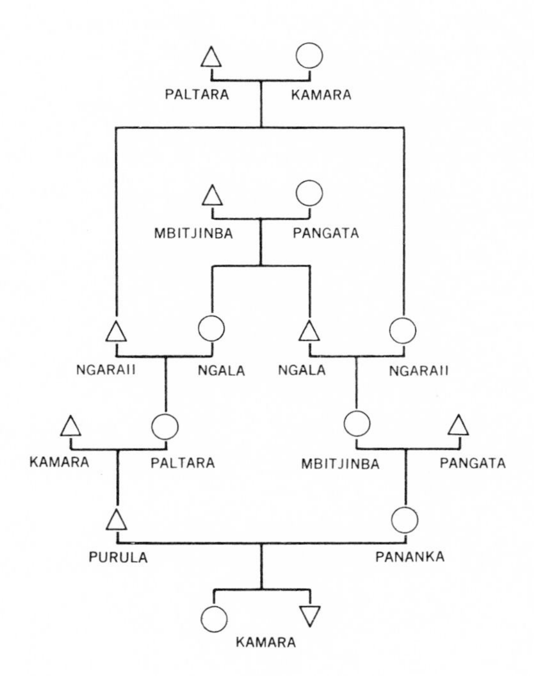

FIGURE 3

The mother's mother's brother's daughter's daughter is the preferred wife. An examination of the relationship terms will show that the members of this constellation are distinguished by separate terms, some of which refer to their position as in-laws rather than to their positions as blood relations.

The key to the Pitjentara system lies in the fact that a man marries his cross-cousin, i.e., his mother's brother's or father's sister's daughter. A man may not marry his daughter,[5] niece, or sister. The only proper person for him to marry is his cross-cousin.

The genealogy of Moita, an old woman of a similarly organized tribe, shows how this system functions (see Figure 4). Here, the relationship terms have been used. Both the father and the father's brother are called by the same term (*mama*), which means "father." The father's sister, however, is called by a term (*minkai*) which signifies her relationship to Moita as a mother-in-law, rather than as a blood relation. The same holds true for her mother's brother. Moita's mother-in-law is her father's real sister, and her father-in-law is her mother's real brother. Her husband is her real cross-cousin on both the maternal and paternal sides.

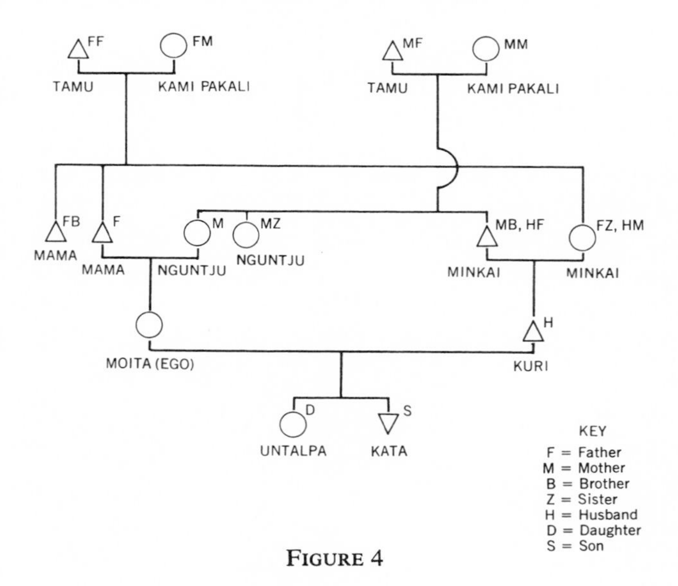

FIGURE 4

In all the small tribes which have an endogamic two-class system, cross-cousin marriage, in the classificatory sense of the term, is considered "proper." In some of these tribes the marriage of real cross-cousins is the orthodox marriage, while others consider the marriage of real cross-cousins somewhat incestuous. In the latter group of tribes, a marriage with the classificatory sister of the real cross-cousin is the preferred form.

We have thus far discussed the functioning of the eight-, four-, and two-class systems. The central and southern sections of the Aranda tribe have a four-class system which really functions like the true eight-class arrangement. The Matuntara group of Lurittya-speaking tribes have an eight-class system. The Pitjentara have a two-class system which is beginning to function like a four-class system. The Mularatara organization is transitional between a two- and a four-class system.

Adherence to the Incest Taboos and the Punishment of Transgressors

A question arises: How closely do the Australians adhere to these rigid rules regarding marriage? Several stories describe in detail what happens to those who transgress the incest taboos.

Old Renana at the Mission told me about a man at Ellery's Creek who had intercourse with every woman he met, even his mother.[6] He had intercourse with his real mother-in-law, his daughter, his brother's daughter, his father's sister, and his niece. The fact that some of these women were his real relations made no difference to him. He was a very strong and audacious man who was a great *tnaputa* (fornicator).

His legal wife was of the right class, but he lived with his three biological daughters as wives. He continued to behave in this manner for a long time. The men "growled" (talked the matter over, expressed their resentment) in every camp, until finally public opinion was transformed into action. Groups of warriors came from Alice Springs and Owen

Springs to reinforce the local men, with Inangimpera of Ellery's Creek as their leader. All the warriors met at Uratnanga. The guilty man saw that danger was approaching and stayed away from the camp. Inangimpera called all the women together and told them to camp separately in the bush, planning to catch the man if he attempted to rape one of them. The woman was to hold him until the men came to kill him.

The women went separately to get seeds. The man surprised one of them in the bush, but when he attempted to have intercourse with her she held him with her arms and legs and cried out for help. The other women, who had been hiding nearby, came to rescue her and attacked the defenseless man with their sharp yam sticks. They put his eyes out and poked the heavy sticks through his body, head, and nostrils until he was dead.

The hair was cut from the corpse and a girdle made of it, as was the custom. This girdle was given to his widow, who gave it to his uncle. However, no blood feud was started. It is hardly conceivable that one should have taken place, considering the circumstances.

This famous case occurred in the time of old Renana's grandparents (1880–1890). He went on to explain that there are various degrees of wrong marriage.[7] Sexual intercourse with the mother-in-law is bad, with the daughter, niece, and sister is worse, and with the mother, worst of all.

Marriage with the classificatory cross-cousin, while not preferred, is acceptable. It is only when a marriage takes place between real cross-cousins that trouble arises. If cross-cousin marriage in the classificatory sense were permitted, then a Purula man could marry not only Penanka, but also Ngaraii and Ngala women. The contradiction between Renana's statement and the rules defined by the class system is the one that always exists between what is required by law or morality and what is done in actual practice.

If we compare the Purula-Ngaraii and Purula-Ngala marriages with the table (p. 7) showing the prescribed marriages of the four- and eight-class system, we see that, from the point of view of the eight-class system, a Purula man, in marrying a Ngaraii woman, is acting as though he were a Ngala. He acts as if the system were four-class, and Purula and Ngala formed one class (Purula). When a Purula marries a Ngala, he is acting as though he were a Ngaraii, i.e., his own cross-cousin. This situation can be explained by the fact that cross-cousins in many of the Australian tribes have sexual rights to the same women.

Aldinga, although a member of the eight-class Ngaratara, married his cross-cousin. He was a Paltara and his wife was a Mbitjinba, not, as she ought to have been, a Kamara. She was not promised to him. He "stole" her and dragged her away to Owen Springs. No one raised any serious objections to this marriage. Their children were Ngaraii, as they would have been had he married a woman of the right class.

Another famous case of incest is that of a man called Wantapari. One day, while hunting, he saw the tracks of Wilti-Wana. She had gone in search of *tjapa* (a kind of witchetty grub). Although she was his mother-in-law, he followed her track until he caught her and forced her to have intercourse. This was *mbanja* (a type of marriage by rape). He took her with him to Annapiti and from there to Atitjani, where she escaped from him. When he followed her to the camp, the people beat him with sticks. He then found another woman who was his daughter; she was a young girl and had not been "opened" yet. When he had intercourse with her, she screamed with pain. He wandered about with her for a time until he came to a camp where he fell in love with another woman. She was his sister, since they were both Paltara. He had intercourse with her in the camp and no one interfered, but he left her when he found a woman of the right class. After a series of short-lived incestuous affairs, he took another woman who was a Ngaraii, his classificatory mother.

At this point Mulda, who was telling me this tale, appeared quite shocked. Until this time his attitude had been one of mild disapproval tinged with secret admiration.

Wantapari was unlike the average Australian who, though he lives a nomadic life, always takes his woman with him. Wantapari was like a sailor with a different girl in every port. When he left his classificatory mother, he met his first wife again and spent some time with her. The people of the camp in which they stayed ignored them. Only Wantapari's uncle, Kurikuri, would give them any meat. By doing this he gave his support to the marriage, or, at the very least, he broke through the wall of social boycott which encompassed the couple. He warned his nephew that the people might take action against him. Accordingly, Wantapari left his wife, and once more began his wandering from woman to woman.

At this point Mulda's narrative was interrupted by someone who exclaimed: "Why didn't you catch him and cut his testicles out?" This exclamation testifies to the universal nature of castration as the talio punishment for incest. Mulda had become so excited that he was pantomiming how Wantapari seized the women and raped them.

Wantapari finally arrived at Nyingukona in central Aranda territory, where all the men in the area joined together and killed him. No hair girdle was made; the body was simply thrown away.

Another similar story was reported at Owen Springs. The name of the guilty man was made taboo, and has now passed into oblivion. They call him Tjirke-wara, which means "lumps of flesh." Lumps are supposed to appear in the flesh of those who have committed incest. He raped the women when they went to get bulbs, and had intercourse with his classificatory mother-in-law, niece, and sister. The relationship made no difference to him, since he also had intercourse with his real mother, real sister, and real daughter. The old men of the area lured him into the bush by promising to show him some totemic ceremonies. When he got there the old men were singing:

> May his liver stick together inside
> And come out of his body like breasts.

This was, of course, not a totemic song, but black magic directed against Tjirke-wara. They caught him and cut his throat. His body was cut into little pieces, and each piece was hung on a different tree.

Maliki, and old Matuntara, told me that the really wrong marriages are those with the mother-in-law and the daughter. Some old men have intercourse with their biological daughters. If it comes to the attention of the tribe, the girl's brothers (both real and classificatory) try to kill the old man. A man whom Maliki knew, Kulto-ngaiatata, roamed through the land with his real daughter. For this behavior he had been speared in the leg several times, but even such drastic treatment did not deter him. He would disappear into the bush before anyone could do him further harm. Finally he was caught at Ltjulka while he was cooking lizards and was speared and killed. His killers ate the lizards that he had cooked.

Yirramba told me of a friend of his who "gave" his wife to a white man. Since he received what he regarded as sufficient compensation, he had no objections to her having intercourse with the white man. However, there was a young man living at the station who was a Ngala. Since the woman was a Kamara, she was his classificatory mother-in-law. In the evening they would disappear into the bush and have intercourse. The affair could not be kept secret. Word was sent to the husband, who came from Alice Springs to Woodford

Well with his son and nephew. They arrived when the wife and the young man were cohabiting. Instead of striking them dead, as might be expected, the husband, son, and nephew sat down and waited until the couple had finished. Considering that the opportunity was favorable and that the Australians are an impulsive people, this was indeed odd behavior.

The three men and the wife then started back to Alice Springs. The Ngala man followed them, waiting for an opportunity to steal off with the woman. He became so bold that he joined the party when they sat down to eat. But later, when they all made a camp for the night, the son and the nephew speared the young man. He ran away with the spears still sticking out of his body, but they caught him and cut him to pieces with their knives; the pieces were thrown into a creek.

A Pitjentara man, Wanka-kakul, lived with his wife and her daughter by her first marriage. He cohabited with both women. After a while he left his wife and went about with his stepdaughter alone. She bore him a child, but it was killed, and Wanka-kakul was speared in the leg several times. Nonetheless he remained with his stepdaughter. We can surmise what would have happened had she been his own daughter.

There is only one female parallel to these stories. There was a woman called Nanunka (Nana means erection; Nanunka means always in search of an erection) who went from Temple Downs to Ndaria to Henbury and had intercourse with every man she met along the way, whether he was of the right class or not. Two men, her uncles, ultimately caught her and killed her. They put her corpse into a dead bullock which had been cut open, and left the two to rot together.

As is evident from the stories just related, infringements of the eight-class rules belong to two utterly different categories. One calls for mild social censure, while the other eventually leads to a tragic dénouement. A northern Aranda informant, in classifying the "wrong" marriages, arranged them in the following order: (1) father and daughter; (2) Wuna (father's sister); (3) niece (some marry their niece); (4) Ipmana (mother's mother and daughter's daughter; marriage with the Ipmana takes place frequently); (5) cross-cousin. My informant finally decided that the marriages between cross-cousins and Ipmana are almost all right.

Another informant called the latter two types of marriage "half-right." They are so frequent that rules exist for classifying the children born of such unions. The rules of the Aranda tribe differ from those of the Lurittya-speaking tribes. The Aranda rules for determining the class of the children of a cross-cousin marriage are as follows:

If a:	marries a:	the children are:
Purula	Ngaraii	Kamara
Kamara	Pangata	Purula
Ngala	Pananka	Mbitjinba
Mbitjinba	Paltara	Ngala
Pananka	Ngala	Pangata
Paltara	Mbitjinba	Ngaraii
Ngaraii	Purula	Paltara
Pangata	Kamara	Pananka

If the above is compared with the table on p. 7, it can be seen that the class of the woman has no effect on the class of the child. The fiction is that the man has married a woman of the right class and descent is reckoned accordingly. The return of the paternal grandfather's class in the grandson is maintained.

The Lurittya rules adhere to the opposite principle. Among them, it is the class of the woman that determines the class of the child. The Lurittya say: "Never mind the man. Call the child according to the mother, for she carried him in her womb." The Aranda, champions of male superiority, express their views as follows: "Call him according to the father, for how can a woman cohabit with herself? With her heels, perhaps? The child came out of the man." Strehlow believed that cross-cousin marriages were regarded as completely wrong, while John Mathew[8] calls them "alternative marriages." That the natives do not regard them as completely acceptable and desirable is clear, but it is also evident that such marriages are frequent and not regarded in the same light as the deeds of a Tjirka-wara.

I know of no case in which someone who married his cross-cousin or his Ipmana was punished or killed, as in the cases of real incest mentioned in the stories. Marriages of this type are simply not proper, but if such a marriage does take place, society ignores it. Society reacts in a very different way to cases of real incest. The desert is a sanctuary for the lawbreaker if he is a strong man, willing to rely on his strength and prowess as a hunter and to give up the advantages of living with society. It is unlikely that he will be attacked by strange natives. Of more importance is his predicament when he no longer shares in the food distribution of his tribe.

Naturally, the Central Australians are aware of the difference between biological father and classificatory father, biological mother and classificatory mother. The horror associated with the idea of intercourse with a classificatory relation is merely an extension of the horror provoked by actual incest (i.e., incest in our sense of the term). The narratives of my informants, their attitudes while telling the stories, and their direct statements left no doubt about this matter. The prototype of all forbidden sexual relationships is that of son and mother, while father-daughter and brother-sister relationships are considered slightly less reprehensible. Cohabitation with classificatory relatives is a weak imitation of the real thing. Save for the existence of a *group* of mothers and a *group* of fathers, the Australian ideas about the seriousness of a type of incest are similar to our own. Having sexual relations with an aunt or an uncle is considered a more serious offense than with a cousin, but far less serious than with a parent or sibling.

When an incestuous relationship begins, society first reacts by ostracizing the couple. If the relationship is continued, the group takes more drastic action. It is only because of incest and ritual offense (which, as we shall see, is the symbolic equivalent of incest) that the group as a whole will take action, for in these instances the group is the offended party.

The Unconscious Determinants of the Kinship System

We have discussed the functioning of the class systems and what happens when the incest rules (a class system is, in actuality, a collection of incest rules) are broken. We must now discover the meaning behind these rigid and encompassing rules.

In Central Australian society we are dealing with a group of human beings whose most powerful motive force is their latent Oedipal striving, and who react to this striving by erecting a series of barriers, none of which is really effective, against the return of the repressed. And so new barriers must continually be erected, or old ones elaborated.

One of the first steps in this process is the creation of a series of mothers, fathers, and so on. This the classificatory system does, and so dilutes the tension arising from the Oedipus Complex, making it easier to handle. We call this tendency toward increasing the number of prohibited objects, in this case women, a phobia. The original object, both desired and anxiety-provoking, radiates these qualities to objects more and more remote, as each of these distant objects becomes identified with the original. Since psychoanalytic anthropology cannot be satisfied by the mere translation of relationship terms, we must attempt to discover the unconscious feelings and rationalizing concepts represented by the terms.

It becomes obvious to the fieldworker that the system is based to a great extent upon the father-son relationship. Two middle-aged Australian men of about the same age visited me one day. While one of the two did all the talking, the other seemed to think it proper to observe a certain modesty and reticence. The first man, Tungala (an Aranda), explained the reasons for this situation. He was the *father* of the other man in the classificatory sense. However, the essential conscious content of his fatherhood was the same as that of a real father. The idea of protection on the father's part and tacit acknowledgment of this protection on the son's part were present. Tungala explained that he had been "taking care" of the other man ever since the latter's father died. Just how he had been "taking care" would be difficult to say. It seemed to consist of assuring the other man and the people with whom they were in contact that he *was* taking care. He gave the other man that infantile feeling of security which, in a society without a classificatory kinship system, might have been lost at the death of the real father.

The relationship between father and son is all-important among the Australian aborigines. The son follows the father when he goes hunting and learns by using him as a model. When the father throws a spear at a kangaroo, the boy does the same with a toy spear. In our society conscious feelings and unconscious ideas and fantasies related to the father are, in Australian society, extended to the classificatory fathers; to the uncle, to the father-in-law (who in many cases is the uncle). Nonetheless, these feelings retain almost all of their original intensity in connection with the real father. On the conscious level, the boy imitates the father in order to gain knowledge of the means of obtaining sustenance. On the unconscious level, the boy desires something which the father owns, and identification is one of the circuitous paths by which this otherwise unobtainable goal can be reached. The object which the father owns, and which is reserved for his use, is the mother, the primal and eternal goal of the child's desires. Another object, owned by the father and desired by the child, is the father's penis. The fact that the father owns the mother and the instrument of pleasure provokes anger in the child.

The aggression of the child came to light in several dreams which I analyzed. In one dream the dreamer appeared as a blood avenger (*leltja*) whose victim was clearly a representation of the father. The child's aggression is projected onto the father and the father is unconsciously aware of it, for that which the child is so desirous of obtaining the father is equally desirous of keeping for himself. Several of my informants told of dreams in which they killed their own children. The concept of the father as destroyer is based both on a correspondence of unconscious attitudes and on the talio mechanism of primary aggression turned against the self. In other words, the father is seen as a destroyer because of the projected hostility of the son and his (the father's) own hostility toward the son who seeks to usurp his place.

The demon (*erintja*), the blood avenger (*leltja*), and the man who uses the pointing bone in dreams, fantasies, and myths are all representatives of the father. This material, furnished by the Oedipal situation and its social manifestations, serves, in its sublimated aspect, as a basis of cooperation between the young and old men of the tribe.

The relationship between father and daughter is less important socially and economically than that between father and son. Its predominating note is an erotically tinged tenderness. Minguri, one of my informants, was seated in my tent telling me myths, while his four- or five-year-old daughter crawled all over him. He muttered indulgently and with a beaming face, "Kunna nurka tara" (vagina with blood), as a term of endearment to the little girl.

A distinctive feature of this erotic father-daughter relationship is that it is acted out more than any other type of incestuous feeling. This type of incest is far more frequent than any other type. The preferred marriage approximates this type of incestuous relationship, for in it the infant girl is promised to an adult man who, while usually younger than the girl's own father, is about the same age as her mother.

In the conception dreams of the women, the unconscious sexual fantasies connected with the father are vividly portrayed. The father appears in the guise of all hostile beings —*leltjas*, mythical serpents, and so on. The large penis of the adult is in itself formidable from the infantile point of view. Desire for the love of the father also involves rivalry with the mother, and so arouses anxiety which is connected with all representatives of the mother-imago. The anxiety-laden aspects of the affects associated with the father on the parts of both sons and daughters are projected onto all distant and unknown tribes, whose members are therefore regarded as *leltjas*, representatives of the "bad father" concept.

Links which connect the son to the mother are less important sociologically than the complex relationship of the son to the father. This is only true, of course, from the somewhat superficial "sociological" point of view, for society is based as much on the reaction formations inherent in the mother-son relationship as on the sublimations and repressions arising from the father-son relationship.

The physical bond between mother and son is so close (as evidenced by the sleeping positions, which we shall deal with later) that society makes corresponding efforts for its dissolution. These efforts manifest themselves in the class system and its increasing emphasis on exogamy, which increases the distance between the mother and her substitutes and the son. Initiation ceremonies and their concomitant castration threat, which strengthens and increases the severity of the superego, effect the separation of mother and son. The emphasis which these primitive societies place on the initiation ceremonies and the phallic *tjurunga* cult unites the males of the tribe with bonds which prevent the Oedipal strivings from reaching consummation. The son has very little in common with the mother after his initiation.

The union between mother and daughter is not nearly so important in these Australian societies as that between son and father, which is emphasized by its superstructure of ritual. The mother has nothing but the arts of everyday life to show her daughter, while the father not only teaches his son these mundane techniques, but also initiates him into the secret lore of the people, thus cementing the bond between them with a series of sublimated primal scenes.

Mother and daughter form a sociological unit as they go about in the bush, gathering edible grubs and roots. This unit often remains in existence after the daughter's marriage. There are three important aspects of the relationship with the mother. She is the first love object; she is the rival in the Oedipal situation; and she is a haven of refuge from the sexual attacks of the male. The female demons in dreams and myths are, of course, her representatives.

Passing from analysis of the parent-child relationships to the relationships that exist between siblings, we will begin with that of brothers. Their relationship is of great sociological importance. They belong to the same marriage class. While they will defend each other in a fight with outsiders, they are more likely to be the antagonists, since they are rivals for the same women. However, in times of serious trouble, save when an incest rule has been broken or a grave ritual offense committed, a man can count on the support of his real and classificatory brothers (i.e., the members of his marriage class) to defend him. On the other hand, the exchange of women at the *ltala* (a licentious festival) is a constant source of fraternal envy and fights. Since, particularly in the case of classificatory brothers, one may be much older than the other, aggression against the brother is frequently an outlet for hostility originally directed against the father.

In examining the sources of fraternal conflict, we can see the results of the breakdown of the class system, that phobic bulwark against incestual strivings. The infantile situation of brothers vying for the same woman, the mother, is repeated and institutionalized in the marriage class system.

The relationship between sisters is similar to that between brothers.

We have discussed the relationships which exist among members of the immediate family. We shall now go into those relationships, either by blood or marriage, which are important in the social organization of the tribe and which gain significance in our eyes because of the feelings and ideas, originally reserved for the natural parents, that have been transferred to them.

Probably the most important figure outside of the immediate family is the uncle, the mother's brother. In the Pitjentara type of social organization the wife is the cross-cousin, and the uncle is therefore the father-in-law. Although these are not a matrilineal people, we find the same mechanisms of fission and displacement among them as exist among matrilineal groups. The uncle (father-in-law) is in some ways a representative of the father imago. While the more tender aspects of the father-son relationship remain within the family circle, the more hostile (castrating) aspects are displaced onto the uncle. He is the initiator, and it is with his family that the adult sexual life is acted out. He is also a type of benevolent mother who feeds and cares for his nephew.

Before we can discuss the role played by the uncle in the unconscious lives of these people, it is first necessary to discuss his actual social and economic role. Among the Aranda, the children of cross-cousins marry. The child of a Ngala woman is a Paltara. The mother's brother, who is also a Ngala, assumes a benevolent, protective attitude toward the child. He gives him food from time to time and, when the boy marries, continues to make these small donations of food to him and his Kamara wife, who is a classificatory sister of the uncle's own daughter. The child of this union is a Ngaraii, that is, of the right class to be the uncle's wife. In order to repay the uncle for his kindness to them, the couple give the infant girl to him, through the medium of the Ngala grandmother. The baby, in her bark cradle, is handed to the grandmother, who places the cradle between the legs of her brother, the uncle. He is to feed her, and marry her when she grows up. If he does not want her, he gives her to one of his classificatory brothers.

These are the more kindly, maternal aspects of the uncle's position. However, because of his major role in the initiation ceremonies of those people who practice cross-cousin marriage (e.g., the Pitjentara), he is representative of all that is inhibiting and castrating about the father. The father-in-law is the circumciser. During the ceremony he receives the boy on a shield, which is an act of adoption. The boy's food distribution obligations are the same with regard to the uncle as to the father. Thus, emotional, social, and economic elements are closely interwoven.

In a society with a classificatory system, the child has many mothers and fathers. Raymond Firth describes this system among the Tikopians:

> In a Tikopian family a child as soon as it arrives at some understanding of the people by whom it is surrounded, distinguishes two persons in particular, to whom it applies the first words of definite speech which it learns . . . "mother" and "father." . . . By a natural tendency . . . to extend the basic terms from the known to the unknown . . . the terms . . . also come to be applied to other individuals who habitually enter the child's sphere of interest.[9]

Since the child has many fathers, how is the uncle, the mother's brother, so differentiated from the rest, and why is so much affect displaced onto him? While the child calls all males "father," he observes certain differences in the way a particular male behaves toward his mother (or, in distinguishing the father's sister, the manner in which some few of the women, whom he calls "mother," behave toward his father). All other men regard his mother as a possible sexual object, while this one man must avoid her. In societies with a brother-sister taboo this difference in behavior is even more pronounced. But even in societies which lack this taboo, the child has a different unconscious reaction to the mother's brother. The mother's brother is a different kind of father, since he has no potential sexual rights to the mother. His relation to the mother is on the same level of sublimation as the child's must be. In this way the child identifies with the uncle. (This may be the explanation of the "lenient uncle, severe father" pattern found in some societies.)

It is in the nephew's identification with the uncle that many of the roles played by the uncle originate. We have discussed the uncle's role as the representative of the father. Onto the uncle, as the only man who has a sublimated relationship with the mother, is displaced the castration anxiety connected with the child's desire for coitus with the mother. This displacement of feelings, originally related to the father, stems from another source as well. From the infantile point of view, the first categories of people differentiated from the multiplicity of "fathers" and "mothers" must be the mother's brother and the father's sister. Since the taboo against marriage between siblings prevents the "fathers" from marrying the "father's sister" and the "mothers" from marrying the "mother's brother," those two individuals must marry each other. Since the children of any of the "fathers" or any of the "mothers" are the "brothers" and "sisters" of Ego, he must marry his cross-cousin, the only person with whom marriage is not incest. These cross-cousin wives are often obtained through an exchange of sisters. The wife and the sister become identified, and the one from whom the wife was taken, the father-in-law, becomes identified with the father. As was mentioned earlier, the uncle is the initiator. Only in a society with such a simple, easily readable ceremonial structure could we find the father substitute in the role of castrator, and reception into the social fold conditioned by the displacement of feelings from the father to the father-in-law by the substitution of an exogamic for an endogamic love object.

Another role which the uncle plays is that of a kindly, feeding mother. In this society the father plays what we generally consider a maternal role with regard to his wife, since there is usually a great disparity between their ages. The sign language of the Pitjentara gives us a great deal of information about this reversal of roles, with regard to both father and uncle. "Father" is indicated by holding the breast and shaking it with the thumb and first finger. This gesture is obviously symbolic of the concept of mother, for which it is also used. The mother, or rather the nipple, is certainly the first part of the outside world which the child perceives in his environment. Whatever the infant feels toward the father must have been carried over from his feelings about this primary object. Alice Balint [10] has shown how the concept of a good father, a good chief, is really due to the overlaying, in both imagination and actual character formation, of the original concept of male aggression by that of the feminine, the maternal. The gesture signifying "uncle" consists of moving the first and second fingers toward the heart and the breastbone. This gesture also indicates the father's sister, who is of course the mother's brother's wife. The Pitjentara explain this gesture by referring to the travels of the child before birth. The unborn child is thought to go from the ancestral cave into the heart of the uncle; from him it passes to the heart of his wife, the father's sister, and from her it goes to her brother, who conveys it to his wife, the mother of the child. This passage of the unborn child from uncle to aunt is a duplication of parental coitus on a sublimated plane. The uncle, who first receives the child from the ancestors, occupies on a fantasy level the same position as the mother, who receives the child from the father. The uncle is in this way a ritual mother.

Many of the conflicts in the natural family are thus resolved in the person of the uncle. As to the other classificatory and real relatives, it is possible to prove that their social positions and socioeconomic obligations (and the taboos which these positions entail) rest, as does the position of the uncle, on the repression, sublimation, and displacement of the affects and ideas derived from the Oedipal conflict. In order to explain the classificatory system, we need only assume that an incestuous libido exists within the family circle, that these incestuous trends are repressed, and that there is a compulsive mechanism of "series formation." Even when the trends in the immediate family are displaced onto more distant relations, the repression is not successful and the old difficulties arise once more. This breakthrough of repressed feelings leads to the development of many avoidance customs and taboos to circumvent the appearance of anxiety.

Very often a man or woman will not approach a group if a particular individual is a member of it. If it is absolutely necessary for him to do so, he will avert his head and look in another direction. This is called *kerintja* (shame, avoidance). Avoidances are far more stringent with regard to in-laws than with regard to the immediate family. This tends to prove that the Oedipal conflict loses much of its virulence through displacement. A man avoids his mother-in-law rather than his own mother. There can be no doubt as to the libidinal basis of these avoidances. Avoidances which exist within the natural family are first practiced when signs of sexual maturity appear in the boy or girl. The natives explain that the in-law avoidances prevent incest, but this explanation is not applied to similar attitudes within the natural family. The natives say merely that avoidance is practiced within the natural family because it has always been the custom to do so. The different explanations show that the repression has been more successful at the original source of the anxiety, and that there is a tendency for the repressed to break through as the feelings are displaced onto more and more distant objects. The original strivings reappear in the substitute situation, and a dramatized repression is acted out in the avoidance and flight.

The predominant avoidance is between son-in-law and mother-in-law, and the avoidance is greater on the part of the son-in-law. Sometimes old women abuse the shyness of their sons-in-law. I was told of one old woman who would often appear suddenly when her son-in-law was eating. When he ran away, she would sit down and eat the food he had left. Strehlow writes that a man may not speak to or go anywhere near his real mother-in-law in the camp, but if they meet in the bush he may use *ankatja kerintja* (taboo talk) or sign language and converse with her from a distance. Were he to speak to her in public, his body would swell and itch intolerably. The mother-in-law avoidance, as all other avoidances, is most intense with the real mother-in-law, and decreases in severity with classificatory relations. It is least important with those members of a given class who come from distant localities and speak another language or dialect.

To a minor degree, the daughter is *kerintja* to the father. He should not look at her too much. This avoidance begins when her breasts begin to swell. The younger sister is avoided in the same way by the brother, while the older sister is not. A boy starts to avoid his mother when his pubic hair begins to grow.

The cross-cousin relationship confers certain rights and privileges. The cross-cousin is the playfellow in both non-erotic and erotic games. In several dreams which I analyzed, the cross-cousin appeared as the dreamer's double. Erotic joking about ceremonial matters is permitted between cross-cousins. They masturbate one another while discussing the size of the other's penis. Among the Aranda, mutual masturbation can also take place between female cross-cousins.

The psychology of the cross-cousin relationship exemplifies the manner in which the psychic mechanism, on which the classificatory system is based, functions. It is the result of "series formation" and of the fission of the original *imagines* by the partial displacement of libidinal and emotional attachments. The sister is a substitute for the mother, the cross-cousin for the sister, and the wife for the cross-cousin.

An important feature of the classificatory kinship system is its intertribal character. All those within the fold are relatives, and therefore good. Those who are not are evil; they are the demons and the blood avengers, representatives of the projected aggressive and sadistic tendencies of the tribesmen. The prescribed attitude of one tribesman toward another is one of tenderness. However, the bonds of relationship overlap tribal boundaries, although intensity decreases the farther we go from the original family. A "father" or "brother" from a distant tribe, whose language he cannot understand, is certainly less of a father or brother to a native than a member of his own tribe. Identification with a member of the same marriage class is subject to the same diminishing intensity with distance, as is the keeping of avoidance customs.

Kinship and social organization are based, in the final analysis, on the parent-child relationship. Extension of the "mother" and "sister" concepts regulates marriage. The maintenance and elaboration of the infantile respect and fear of the father are the factors which hold the tribe together.

NOTES

1. A. Radcliffe-Brown, *The Andaman Islanders* (Cambridge: Cambridge University Press, 1922).

2. H. Cunow, *Die Verwandtschafts-organisation der Australneger* (Stuttgart: Dietz, 1894).

3. In order to clarify the question of kinship regulations, the editor is grateful to Nicholas Peterson for the following statement:

The problem with Róheim's analysis of the kinship system is that he confuses the class system with the system of kin classification and with marriage rules. He is not, of course, the only person to do this. Since the very beginning of Australian aboriginal studies this confusion has plagued the literature.

The class system (usually referred to as section systems when there are four classes and subsections when there are eight) is only a summary of the system of kin classification. Its main significance is in ritual life. It does not regulate marriage but, because it summarizes some of the main rules which govern choice, a spouse should always fall into a particular class.

For example, if there are 400 people in a four-class tribe, then there will be 100 people to a class, fifty of whom will be women. In no system does one have a claim to all the women in the class from which one obtains a wife. Often a person applies more than one kinship term to people in a class. Róheim mentions this on p. 8, where he indicates that the Aranda distinguish *noa* (potential wife) from cross-cousins in the same class. This leads him to say that although there are only four named classes, it is really an eight-class system, since each class can be divided into two. But even if it is divided into two there are still twenty-five women from whom to choose a wife. No man has a right to marry *any* of these women, since there will be many other men who also have claims on them. He will have strong claims on one or two women, based, most often, on direct genealogical links such as MMBDD (i.e., he can claim his mother's mother's brother's daughter's daughter as a wife). The rules by which he knows what he calls people and has claims on them have nothing to do with the class system but with the genealogical system of kin classification. The class system summarizes some of the rules in use in determining what kinship term is applied. That is, the wife's class is the class of the opposite moiety in which women of one's own and alternate generations fall.

4. Carl Strehlow, *Die Aranda-und Loritjastämme* (Frankfurt: Baer, 1908), vol. 4, pt. 1, p. 76.

5. Unless specifically stated, all relationship terms refer to classificatory, as well as real, relations.

6. As noted, a classificatory, rather than real, relationship.

7. Renana was a member of an eight-class group.

8. John Mathew, *Eaglehawk and Crow: A Study of the Australian Aborigines* (London: David Nutt, 1899).

9. Raymond Firth, "Marriage and the Classificatory System of Relationship," *Journal of the Royal Anthropological Institute* 60 (1930): 240–241.

10. Alice Balint, "Der Familienvater," *Imago* 9 (1923): 292–304.

2

THE FOOD QUEST

The following is the beginning of a typical myth, related to me by an old, blind member of the Yam totem who lived on Mission rations:

> The ancestor of the Yam totem was at Ipita-kungu. He killed a euro, brought it back to his camp, roasted it, cut it up with a knife, ate it, and went out again. The next morning, he killed another euro. He brought it and some yams back to his camp. After having cleaned both the meat and the yams, he left them on the sand and went out again. He found another euro, killed it and brought it back to the place where the yams and the meat were lying. Then he moved to another camp. He got another euro, brought it back to the camp and put it down beside the yams. Then he went back again, with the wind blowing towards him, found another euro and killed it. Another member of the Yam totem, a brother of his, met him and the two went together into a good wind. They came back to Ipita-kungu and then they went a little further away in the direction of the wind, and afterwards came back to Ipita-kungu.

The days when he would hunt euros and find yams like his mythical ancestors were the happiest days of the old man's life. It is therefore easy to understand why he elaborated this part of the myth. I did not even copy down everything he said; there were many more euros and yams in the original narrative.

As the myths related by the informants show the food quest and hence the emotional aspect of the almost chronic anxiety about their subsistence is possibly the most outstanding theme; no wonder that the continued struggle for sustenance leaves its mark on the unconscious and hence is also the main subject of dreams, fantasies, and all small talk. This is monotonous for the outsider, but supremely interesting for the natives.

Everyday life consisted nearly exclusively of the food quest. All the activities of the men, women, and children were centered around this one object, the never-ending search for food. A Central Australian native never tired of detailing the events of the food quest, since all his energies were bound up in this search. They did not know the art of storing food. Leisure lasted only as long as it took for a meal to be digested. Conditions were somewhat easier during the rainy season than during times of drought. When there had been a heavy rainfall, as many as twenty or thirty people might camp at one place, making it possible for some to remain at home preparing string, making tools, or getting ready for ceremonies, while the others gathered food. But when we were in Central Australia the drought was so excessive that twenty or thirty people could be brought together only if we supplied extra rations. Conditions were so bad that people wandered about the desert searching for food in groups of four and five. The smaller the group, the more exact the correspondence between the terms "food quest" and "everyday life."

The dreams and fantasies which my informants related to me all showed the central importance of the food quest. Nyiki told a dream in which he killed a lizard, cut off its head, and ate the rest. Then he came back to his camp, went out again, killed a kangaroo, and then swam in the creek. Urkalarkiraka reminisced about her childhood. One time, when there had been a heavy rainfall, many women came to visit her mother. They played, swam in the creek, and had a good time. When she and her aunt went walking in the bush, they met a man who gave them some fat kangaroo meat. Although they both ate huge quantities of the food, there was so much left that her aunt was able to bring home a full bark vessel. Urkalarkiraka imitated the way her aunt walked with the *pitji* (bark vessel) under her elbow. The way she mimicked her aunt made it appear that the *pitji* was an extension of her breast. She then pointed to her own breast to show me how large it had been when she had eaten all that food.

Memories of swimming are always related to plentiful rain and flooded rivers, which means abundance in the desert. Most of Urkalarkiraka's conversations with me centered about the time in her childhood when there had been heavy rains. There were many water holes in which she could swim. And how many ducks! What a quantity of eggs! She even saw a tree full of wild oranges, which is an event in a life centered around the food quest.

Ilpaltalaka, in speaking about her childhood, told of the seeds she ate when she was wandering about with her mother and her grandmother. She gave me a detailed account of how fish were caught with porcupine grass, and how once, when they had already caught many fish, they also found witchetty grubs.

Even when relating dreams, the Aranda maintained the style of the mythological narrative. The following is a verbatim account of one of Yirramba's dreams:

> I saw a great many *alknarintja* and bush women sitting down. They wanted me to come near them, but I would not. There were three groups of them. I dug yams out of the ground,

> while some of the women killed a rat. Some of the women procured bush seed, baked it in ashes and ate it. Then all three groups came back to the camp. An old man (my father) sat in the center of the camp on a big flat stone called "Navel." The women decorated his head with rat tails and then went in search of more rats. The old man sat on the stone all day performing ceremonies.
>
> In the evening, the *alknarintja* women came back. Then they all went away again. When they came back, they made a fire, and drove rats into it. They had caught many rats and they decorated themselves with the tails. They went out again and caught more rats. They cooked them in the camp where the old man was. While one group went out and caught more rats, the other group rested. Then the other group went out to catch rats while the first group rested. They made a fire with a great deal of smoke and gave the rats to the old man. I was making string, some of the women were making circular headdresses, and others were decorating themselves with rat tails. [This was repeated several times.] They drove all the rats into the fire and gave them to the old man.

The rest of the dream does not interest us here, since we are not analyzing the dream in order to discover its latent elements. We are interested instead in the manifest dream text. It is obvious that the old man did not tell the dream exactly as he dreamt it; since it is inconceivable to dream fifteen times of chasing rats into a fire. The elaboration is a result of the constant concern with the food quest.

The Euro and the Kangaroo

Since everyday life means almost the same thing as the food quest, a description of a typical day will serve to show various techniques of food gathering and the subsistence pattern.

Early in the morning the men repaired their spears. The hooks were tied to the shaft of the spears with tendons; those spears which lacked hooks were straightened in hot ashes. Then the hunter was ready to start out. He looked about, shading his eyes from the sun, until he caught sight of a kangaroo. There it was, lying in the shade of a tree! The hunter took careful note of the wind. It must be blowing from the prey to the hunter, otherwise the scent will reach the animal and give warning. The hunter also must take care to approach the animal from behind and to stay under cover of stones and bushes. If the animal catches sight of him, the hunter must retreat to his starting point and begin his approach again. The hunter would creep up as fast as he could, throwing his spear when he was about ten yards from the animal. Thirty yards was about the longest throw a native would risk. These same methods of creeping up on prey were used by the Mission natives who had guns. The natives were not generally very good shots, but they shot from so close that it was rare that they missed their target. At times they were able to come so close to the animal that they could grab the kangaroo by the tail and hit it on the neck with a stick or a stone.

An animal speared in the back will die, but if the spear gets stuck in the ribs the animal lives to run away. If the animal was only wounded by the spear, the hunters followed it. The euro cannot be tracked in the ordinary way, since it runs into the rocky hills where it does not leave footprints. The wounded animal was tracked by following drops of its blood, with the hunters pursuing at a distance until the animal died of exhaustion.

When the animal finally died, the hunters took it by the forepaws and dragged it to a shady spot. Then came the *otu tunanyi*, which is a Mularatara term meaning "breaking." The kangaroo was placed on its back and its forepaw placed over the arm of the hunter. The foreleg was broken and the marrow bone removed. This bone was thrust into the animal's flesh because the natives believed that this would prevent the animal from becoming stiff. They thought that when the flesh became stiff, its weight increased, thus making the animal too heavy for them to carry home.

The animal was then turned so that it lay on its side, the belly slit, and the gut removed. At this point the men drank some of the kangaroo's blood. Kanakana, one of my informants, said: "It tastes sweet like human blood and gives us strength." The hunter squeezed the feces out and placed the gut in the stomach. A stick was pushed through the bag-like stomach and the animal was carried in this manner. A hill kangaroo (euro) was carried on the head, all other kangaroos were carried on the shoulders.

If the hunter passed a pool on his way back to the camp, he would throw the animal off his shoulder and stop for a drink. There was no hurry about returning. Before drinking, the hunter rinsed his mouth with water. He might vomit from the great heat if he did not do this first. He splashed himself and then rested in the shadow of the trees that surrounded the water hole.

When he felt rested, he dug a long trench which he lined with dry brush and grass. Then, standing on the buttocks of the kangaroo, he broke the hindlegs at the ankle. He pushed a stick into the flesh, removed the tendons that he needed for string, twisted these tendons around the stick, and put them away for safekeeping. He lit the dry brush and, holding the kangaroo by its fore and hindlegs, swung it in the fire in order to singe off the hair.

When the hair was completely burned off, the *pubes* were cleaned. Then the legs were broken off completely and placed beside the fire, as was the tail. The body, legs, and tail were covered with hot ashes, and the gut was also baked.

The fat of the gut and the gut itself were given to the old men at the camp. Anyone who helped the man who actually killed the animal also received some of the fat and the gut to give to his father in case of his absence. The hunting party sat around the fire until the meat was cooked, at which time it was laid to cool on boughs placed beside the fire for that purpose.

When the meat was ready, the back was slit open. One of the men raised the body with a stick thrust into the rectum, while another broke off the ribs with his yam stick. A small hole was made to loosen the hip bones on either side. The ribs, the hip bones, and the backbone were cut out of the flesh, and the backbone was broken in the middle. The neck was cut with great care lest the Adam's Apple be destroyed. It was believed that anyone who destroyed the Adam's Apple would have his own destroyed in the same manner. The neck was then broken off the backbone.

When all this had been done, the parts were laid out in a prescribed manner. In two equal rows were placed the two rib parts, the two legs, the upper part of the backbone, the lower part, the stomach, the neck, the head, and the hip bones. The tail and the tail end of the backbone were never taken by the man who killed the kangaroo, for he would never again be lucky in hunting if he ate them.

If the hunter had been alone when the kill was made, he cut the body into two halves, the upper part called *kultu*, and the portion from the hip down called *karil*. He tied the two parts together with a string and carried them to his father, who then divided them as described above.

Kangaroo and emu were the big game of the desert country. Bringing one of these animals into the camp was an occasion for a holiday. Considering the central significance of their quest for subsistence and the underlying anxiety, it was an eminently socioeconomic affair, conducted with fixed rules of distribution and an almost ceremonial and prescribed formality. The smaller the game and the poorer the quality of the flesh, the more it became a private, nonsocial matter. If the hunters had been lucky, the day ended with a feast, with the best and most fatty parts of the animal being given to the old men. If the hunters had not been lucky, the people ate whatever grubs and roots the women and children had been able to find. Except for these special instances, one day was much like another.

This description of the hunting of the hill kangaroo relates only one of the many ways in which food was obtained. The narrative of Aliumba, a Ngaratara woman, told of another method of hunting the kangaroo, as well as providing a fairly accurate portrayal of everyday life in the desert:

> When I was a little girl, there were a great many people at Maituka. My father, brother, and mother went to Wanaraka and found a great deal of *quandong* there. While they were there, Atalkunta raped (*mbanja*) Kwita. He also had a fight with his son because he had not given him [Atalkunta] his share of meat. The tail, backbone, gut, and the fat of the gut must go to the father or father-in-law. After the fight, we got many kangaroos. The women made a big noise and drove the kangaroos into the ambush where the men lay in wait with their spears.

The ambush method of hunting was used for both the kangaroo and the euro. The men built a semicircular fence and lay in wait behind it all night. This was employed only on a night when there was a great deal of moonlight, for otherwise the men could not see well enough to take aim. The fence was built near a water hole which the animals frequented. The euro makes a hissing noise when drinking, while the kangaroo makes a blowing noise when it laps up the water. The men threw their spears at the animal, but did not come out from behind the fence to pick it up lest they frighten away other animals approaching the water hole. Animals which had only been wounded and escaped were tracked by the hunters the next morning. If the hunters were lucky, they killed many kangaroos in one night.

Some of the Aranda used another method of trapping the kangaroo. They dug large holes in the paths used by the animals and covered them with dry brushwood. The brushwood was covered with grass, and the grass with sand. The kangaroos which fell into the hole could not get out, and the hunters could kill them at their leisure.

Basedow[1] describes a combined technique in which both the ambush fence and the trap are used. Another method of capturing game involved driving the animals before a prairie fire. The desert bush was ignited in a semicircle, with its open end facing the hunters. The animals fleeing the flames were then speared by the hunters.

The Emu

The animal next in importance to the kangaroo and the euro was the emu. The hunter followed any fresh emu tracks, looking behind every sand hill until he caught sight of the bird. He sneaked up to it, using the same precautions as were employed with the kangaroo. If there were no bushes between hunter and bird, the hunter created an artificial screen by sticking branches into his belt and crawling on his stomach. The emu has excellent eyesight, and the hunter must stay out of its line of vision. The bird was speared under the wing or at the hip.

The hunter took great care not to come too near the emu, for he feared its claws. It was believed that the bird was stronger than the kangaroo, and could thrust the spear into the hunter's stomach when he came to draw it out. The hunter therefore allowed the bird to run until it collapsed. When at last the bird was dead, he removed the spear, tied the bird's legs to its neck, and carried it back to the camp on his head.

The hunter hid the bird behind a bush just before he entered the camp. His manner when he joined the group was that of an unsuccessful hunter, but when the old men saw the blood on his spear they knew that he must have killed something. When they questioned the hunter, he replied: "I got some bad meat and I threw it away behind that bush."

When the tribesmen saw the emu, they were very happy, for to their taste its meat was better than that of the kangaroo. Only the old men of the tribe were allowed to touch the bird. They plucked it, saying: "Oh, how fat he is." The skin with the subcutaneous fat was cut off, and the legs removed. The ground was covered with emu feathers and the legs and body placed on top of them. Another layer of feathers covered the mat, and on top of this they put hot ashes. Above the ashes they placed branches of the tnurunga bush, on which they lay the skin and the fat. The skin was covered with hot sand and feathers, on top of which was placed a thin layer of fat. Then more feathers and ashes, a layer of fat, and finally another layer of feathers and ashes.

The old men ate the gut first. The fat was eaten when it was only half-cooked. Before the skin was placed on the heap, the places where it had been cut to remove it from the bird were sealed with a stick. This stick was removed and saved, for the old men used it to grease their hair. The fat that adhered to the skin was melted and drunk by the old men.

The father-in-law and brother-in-law of the hunter placed the legs on a bough and cut them up into small pieces. The hunter's father touched the body of the emu and said: "Very fat." He then called the father-in-law and brother-in-law to the feast, but did not join them. The old men sat on a mound and divided the meat. One of them gave the shoulder bone to the hunter. He gave the neck to his mother and the sides to his wife. The rest of the meat was shared among the people, but the fat was reserved for the old men.

The emu was also hunted by the ambush method. When the hunters saw many emu tracks leading to a water hole, they made a nest in the bough of a nearby tree. The hunters watched from a sand hill and, when they saw an emu approaching, quickly climbed the tree. While the bird drank, a spear was hurled into his heart. The hunters cooked the bird on the sand hill and waited for the next one to appear.

The Australians also killed the emu with poison. The water in which the leaves of a certain bush are soaked is fatal to the emu. The bush was beaten into little pieces and the pieces thrown into a water hole frequented by the emus. The hunters needed only to sit and watch. A few minutes after a bird drank the water, it was dead. The gut was thrown away, but the flesh was kept, since the poison did not enter it. Among the Nambutji, this method of procuring emu flesh was one of the privileges of old age. Another method which the Katitji permitted only the old men to employ was one in which a hole was dug in one of the emu paths and a sharpened stick placed in it. When the bird fell into the hole, it was impaled and killed.

Other Game Animals

Of the smaller game, the *inarlinga* was especially reserved for the pleasure of the old men. When one was killed, the body was first washed in water and then laid on hot ashes. After it was cooked, the spikes were scraped off with a spear thrower. The old men broke the bones into small pieces and ate the flesh. If the animal's nose bled when it was killed, they recited a short speech to it, asking its forgiveness. They believed that if they neglected to do this, the soul of the *inarlinga* would tell the stones of the hills to make the hunter's toenail come off and cause him to fall when next he hunted the euro. If the incantation was recited in the proper way, the *inarlinga* would admonish the stones not to trip the hunter when he went after the euro.

Hunting the rock-wallaby was a feat of mountaineering. The hunters climbed the high steep cliffs and searched until they spied the small gray animal on the rocks below. It took great skill to climb the cliffs, and the hunters were able to carry only one spear and no spear thrower. The hunters dropped their spears on the animal, or, if they could, stabbed it. If the wallaby was hiding among the rocks, the easiest way to kill it was to drop a stone on it. The hunters either made a fire and ate the animal on the spot, or carried it on their heads to the camp, where it was given to the father, brother, uncle, or father-in-law of the man who actually killed it.

Dogs were frequently used in hunting the sand-wallaby, since they could chase it well in the flat country. The wallaby would run through the grass where the dogs were able to follow it. When its paws were bruised by the stones, it stopped to lick them, and the hunters were able to come close enough to spear it. The *okalpi* is another kind of small wallaby. The hunters hooked it with their spears as it ran through the porcupine grass, and then choked it.

The hunter was frequently assisted by his wife when he went out for the opossum. One of them drove the animal out of its hole and the other hit it on the head with a stick. At other times they followed the tracks of the animal. If they led to a hollow tree, one of them climbed up and hit the animal on the head with a long stick.

The wildcat was important from a mythological as well as nutritional point of view. The Aranda said that there were many wildcats at Ltalaltuma, one of the most important Wildcat totem centers. They described the way in which they hunted the wildcat as *warkuntama,* the word used to describe the way the young men circled around the performers at a totemic ceremony. The hunters ran in ever-decreasing circles around the wildcat. The animal became frightened and sat down, and the hunters continued to circle until they were able to spear the wildcat at close quarters.

The Pitjentara pulled the wildcat out of its hole by its tail. If by chance it was not facing into its hole, the fingers of the hunter were likely to be bitten off. When they had gotten the animal out of its hole, they smashed its head on the ground.

The eagle, whose flesh was not eaten, was caught for its feathers. The hunters hid under a screen or fence after they had placed some meat as bait on the high rocks. When the bird swooped down, the hunters either speared it or threw their boomerangs at it.

When the Central Australians came upon a group of ducks swimming in a water hole, one of the hunters made a great deal of noise. As the ducks flew away, the other hunters threw their boomerangs at them. The same technique was used for the pelican, the shell parrot, and all kinds of cockatoos.

Lizards and snakes were by no means despised as food, since good food and fat food were synonymous. Good tracking was essential in hunting them. The animal was tracked to its hole, pulled out with the hunters' fingers, and then either speared or hit on the neck with a stick. The hunters used great care lest they spoil the fat by bruising the body. One nonpoisonous snake, the *kunia,* was an important item on the Central Australian menu. The hunters tried to catch it before it could go into its hole, for its nest was so far underground that it was too much trouble to dig for it. They employed an archaic method of dealing with the snake once they had caught it—they bent back the neck and bit off its head, explaining that they did this because they were sorry for the snake and did not want to hit it with a stick since it "had a big *tjurunga*." The mythological importance of a totem having anything to do with the attitude adopted toward its animal representative was not a frequent occurrence in Central Australia. I think that many regarded the act of biting off the snake's head as a feat of courage.

Women and the Search for Food

The life of the women was as importantly bound up in the food quest as was the life of the men. The association between the male and animal life and the female and plant life is a feature of many primitive societies. Early man was a hunter, his female companion a food gatherer. He was concerned with the fauna, both in everyday life and in ritual; she was concerned with the flora. Besides the obvious practical reasons for this division of labor, there is also a deep-lying biological contrast upon which this separation rests. Plant life is concerned with storing up energy, animal life with consuming it. Biologically speaking, plant life is anabolic, animal life catabolic. Certain biologists, limiting their attention in the main to the lower forms of life, have plausibly maintained that males are more catabolic than females, and that maleness is the product of influences tending to produce a more catabolic way of life. If this assumption is correct, maleness and femaleness are merely a repetition of the contrast between the animal and the plant.[2]

The life of the Australian woman centered around the vegetable world: berries, fruits, roots, seeds, bulbs, and so on. The various kinds of yams made up the major part of the vegetable diet. The bulbs were dug up with a yam stick, a tool as typical of the women as the spear was of the men.

One hand shoved the yam stick into the sand while the other scooped the sand out of the hole. The yams, cooked on hot ashes, were rubbed between the hands to remove the skin before they were eaten. The smallest variety of yam, the *nyiri*, was cooked on hot ashes in a bark vessel.

Seeds were another important item in the diet. The method of gathering and preparing them was more or less the same for all types. A basket was dipped into the grass and the seeds swept into it off the tops of the stalks. When the basket was full, the women emptied the seeds into a round hole which they had dug in the hard ground. Then they leaned on a bent stick and jumped up and down on the seeds with their bare feet. The husks were blown away, and the clean seed that remained was placed on a grinding stone and mixed with water. Beneath the grinding stone, which had a hole in the middle, was a gourd resting on hot ashes. As the seeds were ground, they fell into the gourd and were roasted with a fire stick. Some seeds were made into cakes, while others were just mixed with water and eaten with the fingers.

One small representative of the animal world belonged to the sphere of activity of women and children. This was the witchetty grub, the larva of the big Cossus moth. It was regarded as a great delicacy by the natives and eaten by most Europeans who had lived in the bush for some time. The taste resembles that of scrambled eggs, but is considerably richer. Children often spend the better part of the day digging for these grubs. When Mrs. Róheim offered little Aldinga some strawberry jam, he exclaimed, "Maku" (witchetty), partly because the shape of the preserved strawberry reminded him of the grub, but mainly because of his delight.

The children ate the grubs raw, or slightly roasted. Tjintjewara described the manner of obtaining and preparing *maku*. When the children saw a particular kind of bush, they examined its base to see if there were any of the shells which the moth discards when it leaves the chrysalis. If these shells were present, the children knew there were grubs in the bush. They dug into the ground until they found the root of the bush, where they saw the excrements of the *maku*. They then broke the thick root with their yam sticks and inside it found the much coveted delicacy. Strehlow presents a long list of grubs and the bushes under which they were found.

The Central Australians are very fond of wild honey. The wild bees deposit their honey in hollow eucalyptus trees.[3] A myth of the Honey Ant totem which I obtained from Rungurkna, an Aranda from Burt Plain, illustrates the natural history element in myths, and tells us something about honey as a foodstuff:

> Two ancestors of the Honey Ant totem lived at Ulpiramba (Hollow tree). One of them was called Irpilanuka and the other, Ultamba. Irpilanuka was an old man while Ultamba was a young boy. There were also two women in the camp called Iwinjaka and Ndowala. The young boy camped downstream from the old man. When the boy brought honey to the old man it was refused. The honey had been taken out of a hollow tree with a stone and then heated. When the two women brought honey to the old man he again refused to take it. He said that it was too hot. The next day, the boy did not come home. The two women again went to get honey for the old man, which he again refused because it was "too hot." The old man then decided to get his own honey. He had a big *illapa*[4] while the boy and the women had only stones. When the boy returned once more with the honey bag, the old man shoved him away and showed the boy the clean honey which he had gathered.

Food Taboos

Almost all the food taboos among the Central Australians related to the young boys and girls. The explanation offered as to why a certain plant or animal was taboo was always that some part of the child's anatomy would become like the article of food, or that ingestion of this food would interfere with some bodily function. Most often it was the genitalia or some secondary sexual characteristic which would be injured, or some sexual function which would be disturbed.

Certain foods were taboo to the young at all times, and others only at certain critical periods of development. Among the Ngaratara, various plants and animals were taboo for the following reasons: the boys would bleed too much when they were subincised and the girls would menstruate too profusely; the whiskers and breasts would not grow; the penis would become stiff and hard and subincision would not be possible, or possible only with great difficulty and pain; the young people would cease to grow.

The Aranda used the same term for taboo food as they used for the avoidance customs—*kerintja* (shame). Their taboos were similar to those of the Ngaratara. The Pindupi and Yumu groups had certain foods which were taboo for pregnant women and their husbands. If taboo food was eaten by the young, in many instances the punishment was a sore penis after subincision. Among these people, all types of snakes were taboo for the young. The Pitjentara had similar taboos.

In all the ethnological areas which I know, one food taboo is usually given precedence over all the others. Among the Central Australians this animal was the *inarlinga*, the spiny ant eater. The punishment for eating it was always related to the penis or to subincision and, in the case of the girls, to the clitoris. The *inarlinga* played a prominent role in Aranda mythology. He was one of the ancestors who initiated the young men, but he did not perform the circumcision in the correct way. He cut the penis off, and the young men died as a consequence of the operation. This infuriated the other ancestors, who threw their spears at the *inarlinga*. The spears stuck in his body, and that is how the ant eater got his spines. It is apparent that the castration complex is symbolized by the *inarlinga*.

Old Yirramba told me that another name for the *inarlinga* was *tuanyiraka*. This was also the esoteric name for the mysteries of the *namatuna*, the small bull-roarer received by the initiate after circumcision. The large *tjurunga*, which the boy received after subincision, and which was considered a "second self," was also known by this name. The *inarlinga* was believed to have a *kuruna*, a soul, which was the real *tuanyiraka*. When the first *inarlinga* was speared by the other ancestors, the soul came out and made a humming noise like a bull-roarer.

If the *inarlinga* represents the bull-roarer (the *tjurunga*), then it is a representative of the ancestors, the old men, the fathers. The other ancestors, the ones who speared the *inarlinga* for castrating and killing the young men, were the earliest sons. The original castration anxiety remains. Were a Central Australian boy to kill and eat the *inarlinga*, something would happen to his penis. The power of the old men had its source in the castration anxiety of the child.

Superficial investigation of the character of these taboos reveals only the automatic nature of the punishment. If we translate this into psychoanalytic terminology, we can say that there is a mental representation of the parents which has superego qualities. There were, however, some areas in which this internalization was not complete, in which the punishment did not follow automatically. If a man did not give the specified portions of his kill to his father and father-in-law at the food distribution, he was subject to a special kind of evil magic which was placed in his food by the old men.

Punishment for violation of the food taboos was always

directed against the genitalia or an area directly related to them, and the taboos were directly related to the initiation rites and to castration anxiety.

The Medical Arts of the Central Australians

Since the women had the greatest contact with the plant life of the vicinity, it was they who practiced that part of the medical arts which employed plants and herbs. The art of medical magic was chiefly in the hands of the men. A solution of sweet grass was drunk as a specific against colds, as was a solution of kangaroo grass. The leaves of the arata bush were dried, powdered, and mixed with water; this also provided relief from colds. All of these medicines were used by Urkalarkiraka, my informant about all things medical, during the great influenza epidemic which decimated the Aranda at Alice Springs. When I was in Central Australia, tea and sugar were being substituted for these older remedies.

The leaves of the pine tree were dried and then soaked in water. When splashed over the entire body, this mixture cured headaches. A decoction made of the aratna ulqua bush was splashed on the body to relieve all kinds of pain. The Aranda made a bed and a pillow of *indjipinjaatjina*[5] and smelled it. This was supposed to cure the common cold.

The nest of the *iwupa,* an ugly worm with a poisonous sting, was placed on sores in the belief that the sores would first swell, "get hot," and then heal. Only two-thirds of this belief seems to have been borne out by actual practice. The persistent application of *iwupa's nest* to wounds often annulled the treatment of the Reverend Albrecht at Hermannsburg, but it would seem that experience had no effect on popular medical views. This was the only instance in my experience where medicine of the herbal sort was connected with mythology. The ancestors were said to travel inside the *iwupa,* and it is they who were supposed to draw out the poison and cure the sores.

NOTES

1. Herbert Basedow, *The Australian Aboriginal* (Adelaide: Preece, 1925), pp. 143–144.
2. W. I. Thomas, *Sex and Society* (Chicago: University of Chicago Press, 1907), pp. 3, 4.
3. Basedow's statement that there are no wild bees in Central Australia is erroneous. Cf. Carl Strehlow, *Die Aranda-und Loritjastämme* (Frankfurt: Baer, 1908), vol. 5, p. 5.
4. No translation for this word could be found—*Editor.*
5. No translation for this word could be found—*Editor.*

3

THE OLD MEN AND THE CHIEFS

The Position of the Aged

Before we can discuss the position of the aged among the Australian natives, we must right a few erroneous impressions given by several other writers in the field. The severity of the gerontocratic rule has been greatly exaggerated. Porteus,[1] who spent only two weeks in Central Australia, was ignorant of the language, and did not have the services of an interpreter, wrote of the extreme subservience of the younger members of the tribe. He states that marriages were arranged by the tribal council. I found no evidence which corroborated these statements. Tribal society was democratic. Young and old met on a basis of mutual friendship and good will. Had the young men been so subservient, the fights which were so common between young and old men could not possibly have taken place. Tales of fights which had occurred between father and son were related to me without scandalized comment.

Other anthropologists have referred to the sexual privileges of the old men. Strehlow stated that among the southern Lurittya, old men would frequently cohabit with their daughters after the deaths of their wives. While this undeniably occurred with some frequency, it was the act of an individual transgressor, never condoned by public opinion. The class rules were binding for all the people, and there were no sexual privileges for the old men in this sense. One of my informants, in relating a myth to me, explained the behavior of the hero by saying that he was a big chief and could even marry his mother-in-law if he wished. This statement seemed based on unconscious acknowledgment of the father's omnipotence, rather than on any conscious or social acceptance of deviant behavior on the part of the chiefs or the old men.

The old men did have a certain authority due to their position as chiefs, which, as Strehlow tells us, was that of a universal father. The old men did not permit the young boys to have intercourse before they had been circumcised. Offenders were speared in the leg. However, the youngsters did not necessarily obey the commands of the old men, but merely arranged their affairs in such a way as to avoid detection. Although the old women enjoyed rather less prestige than the old men, they were thought to be in possession of "magical" knowledge and were often called *arakutya knaripata* (woman father). In general, age was venerated in either sex.

This veneration was not carried to the extremes that some writers would have us believe. Adult women would scold and beat their aged mothers for trifles. Children were not expected to respect anything save the ceremonial objects connected with the phallic cult. We used to give the children canned milk in the afternoon. On several occasions their mothers stayed to get a glass for themselves. Some of the older children objected violently to this, and shouted: "*Kunna nurka tara* (bloody vagina). You are taking our milk!"

One afternoon, when we were to be shown a kangaroo ceremony, the men assembled without the usual decorations. Pukuti-wara, the leader of the ceremony, cursed them and told them to get ready immediately, but the group refused to move and sat staring off into space. When they finally rose and began to get ready, it was not Pukuti-wara's anger and imprecations, but my promises, which prevailed. The precepts of respect for the aged did prevail in some situations. Old blind Yirramba always found some younger man willing to lead him by the hand when he wished to go hunting. He would speak of "hunting," when someone else actually did the hunting for him.

Strehlow[2] gives a chart containing the names of almost all of the age grades, which correspond to rank. Many of these names indicate not a period of life, but a phase in the initiation ceremonies. These ceremonies were the backbone of the whole social organization. Although these distinctions existed and conferred rank and status, the rank did not entail domineering and subjugating behavior. Obedience shown the elders was simply an extension of the child's natural inclination to imitate.

Gerontocracy manifested itself chiefly in the ceremonial life and in the food taboos and the rules governing food distribution. The old men taught the ceremonies to the young men in exchange for food. The idea of selling *tjurunga*, or (what amounted to the same thing) performing ceremonies for the white man in exchange for supplies, was entirely in accord with their social system. When I asked to see some ceremonies, the old men told me that I was doing the same thing as the young men of the tribe did when they offered food in exchange for learning the rituals.

When one of my informants, Wapiti, was a young man,

there were many kangaroos and emus where he was living. He gave a great deal of his meat to an old man who was visiting his people, and the old man showed him a ceremony of the Emu totem and of the Wildcat totem. The old men performed the ceremonies and the young men brought them food, either as a payment or as an inducement for the performance.

That power which the old men possessed derived from the castration complex. The workings of this mechanism were particularly evident in connection with the food taboos and the rules of food distribution. The language of the Ngaratara contains another clue to the association between the power of the aged and the castration complex. *Mokunpa tjurunga* means "big *tjurunga*." The word *mokunerama* means "to become angry and excited," and is generally used to describe the reaction of the men when one of the uninitiated or one of the women sees a *tjurunga*. The Ngaratara word for "old man" is *mokunpa*. It is evident that the old men become angry when the *tjurunga* are seen by persons who do not have that right. "To see the *tjurunga*" may have been a euphemistic way of describing sexual intercourse. This was punished by the old men, and the punishment for such activity was always castration.

The fact that the same word meant both "big" and "old" shows that the power of the old men originated in the infantile situation. Only a child considers "big" people and "old" people the same. The old people who were physically bigger than he was during his infancy continue to be "big" and "powerful" even when he is an adult and physically bigger than they. The old men, the fathers, can do something that the child cannot. They can eat (which is the Aranda euphemism for "cohabit with") the mother. The child would like to take his father's place, to eat the food that was reserved for the father. But because he fears the father's wrath descending on his genitals, he does not. The magical powers which the old men were reputed to possess originated in the repressed infantile memories of what the father could do. These acts of the father's seem supernatural to the child because he cannot do likewise.

The Chief

The chief is the universal father, even of the old men. The chief among the Central Australians functioned in the following ways. He convened tribal meetings, acted as chairman, and arranged the ceremonies. He was the custodian of the sacred cave in which the *tjurunga* were kept. Before a boy could be circumcised, his elder brother had to ask the permission of the chief to arrange for the ceremony. In times of war, it was the chief who sent messages to friendly groups asking their participation in the expedition, and it was he who divided the booty.

The rank of chief was inherited through the male line, passing from the eldest brother to the next eldest brother, and so on, to the youngest brother. At his death the title passed to the eldest son of the eldest brother. The exact meaning of the word "chief" will perhaps be made clear by the following example. One of my informants, a man called Qualorka, was the chief of Lalkara, Putatunga, Kapalji, and Paura. When so many place names are mentioned in the name of a medieval baron, we assume that either he or his family owned estates at the places mentioned. Among the Central Australians, no territorial rights were connected with any of the place names in the title of a chief. Not only did the chief in question own no land at these places, but he also had little influence over the people who lived there. Chiefship and the titles associated with it were important only with regard to mythology and ceremonials. One of the place names might refer to the place where a chief and a group of his tribesmen lived. If he were the keeper of the *tjurunga* of this group, then the place name would be included in his title. The other place names have a purely mythological significance. Their meaning came to light in the following way. My informants would often stop in the middle of a myth they were telling me and state that the remainder of the story belonged to the chief of a place mentioned in the myth, and therefore they did not feel they had the right to continue with the tale. Wherever a ceremony was performed, the chief of that place had to be present. Although all the other natives might have known the ritual, he was the only one who had the "right" to know it and to arrange for its performance.

Many individuals were therefore the chief of any given place. Anyone whose myths related to a particular place was, in a sense, a chief of that place. The "real" chief of a place was the keeper of the *tjurunga* there, but in another way, anyone who had been born there, or whose father, mother, grandfather, or uncle, in both the real and the classificatory sense, had been born there, could be regarded as the chief. In the absence of the "real" chief, he had the right to organize and perform the ceremonies connected with the place.

Because of the foregoing factors, every grown man among the Central Australians was the chief of one place or another. Class distinctions, indeed distinctions of every kind, increase as we go up the ladder of civilization. That every man should be a chief, and that the function of being a chief should be only an augmentation of the function of being an elder or a father, is what we would expect in an extremely primitive society. I asked Moses, an old Aranda, about the functions of the *inkata,* the chiefs, and of the *yenkua,* the old men, and the differences between them. He stated that the *inkata* taught the young men to hunt, and in return received gifts of food. The *yenkua* (whom he described as a man so old that he could scarcely walk and who had only skin and no flesh upon his bones), like the *inkata*, could convene the initiation ceremonies. He concluded that there was little difference between the two, except that the *inkata* was the custodian of the local *tjurunga*.

We may conclude that the *inkata* was the universal father. As such, he instructed the tribe, his children, in the art of hunting, and received their homage in the form of food. The *yenkua* represented the uncanny or supernatural aspect of the father. Although he had no physical strength and could hardly move, he was possessed of magical powers for which he was revered and obeyed. The *inkata* represented the phallic, the primal scene father. He was the active man, the hunter in the prime of his life.

Several chiefs whom I met had considerable authority. In general, the authority of such a man was a function of his unique personality rather than of his titles, although in a few instances the latter did play some role in the status of the man and in the deference granted him by the other natives. A man with strength, courage, and a dominant personality received the title of *inkata knarra* or "big chief," while other men with similar titles but more passive personalities were called *inkata kurka* or "little chiefs." We will describe the personalities and status of several Central Australian chiefs. This will serve not only to illustrate the concept of chief among these people, but will also provide a picture of several Central Australian males.

Pukuti-wara was one of the most important chiefs I met. The grandson and heir of the great chief of Malu-piti, the place where the kangaroos were made, he was a stern, rather impressive individual. Once, when we were bringing rations to a group of natives camping near the River Palmer, we came upon him standing directly in our path with his spear in hand. Although he appeared to take no notice of us, our meeting with him did not take place by chance. He was waiting there for us, particularly for the rations we were bringing. His attitude then was merely an exaggeration of the usual stolid demeanor of his people. They were slow to perceive and accept anything new in their lives. Pukuti-wara not only had this trait, but he also had many paranoid tendencies. He spoke constantly of demons and of the doings of his dream soul. One of his testicles was missing; he said it had been removed when he was initiated into the guild of medicine men. This might have been a lie, and the testicle may simply never have descended. Whatever the explanation, it was not surprising that castration played such a major role in his dreams and fantasies.

He seemed to be completely lacking in a sense of humor. He was the only man of all those I interviewed who knew no folk tales, apparently regarding them as beneath his dignity. He claimed that such things were for women and children and not for Pukuti-wara, the custodian of the big *tjurunga*, the famous medicine man, who was himself "half emu." The only person in whom he evinced any interest was his small son, Aldinga. This small, potbellied child clung to his hand and went everywhere with him. My wife first pitied the little boy, and then grew to like him. When Pukuti-wara saw the attention Mrs. Róheim paid to his son, he unbent and became much more amiable.

Pukuti-wara's prestige was founded only in part upon his being the chief of a mythologically important place. Most of his influence was based on his austere and forbidding countenance and his fame as a medicine man. The fact that he had been forced to leave his own group of tribesmen for having killed a man with a spear did not detract from the awe he inspired. He had once been cured of a minor ailment by Mr. Kramer, the Missionary at Alice Springs, and since that time had been a friend of the white settlers. They all sincerely regretted his death, an unusual sentiment on the part of the white settlers, who would hardly have been called the friends of the natives. He had never been guilty of stealing cattle, although he had had many opportunities to do so. He refrained from this crime partly because of his friendship with the settlers and partly because he could not risk the loss of dignity involved in going to jail.

Uran-tukuti was the chief of Ilpila, a place as important mythologically as Malu-piti. Since his power as a magician was considerably less than that of Pukuti-wara, and he was a more retiring man, his influence was not nearly as great as that of Pukuti-wara. He was the chief only so long as there was no older man present to represent his group. What little respect he was accorded was due to his title; his own personality added little to his status.

The position of Yirramba was quite different from that of the other two men. His status when I met him was not the same as it would have been had the Aranda been a free and independent tribe, and Yirramba's authority not detached from its local and totemic sources. In Hermannsburg he was regarded by the other natives as an alien and intruder. His personality was not such as to increase the little influence he did have. He was a crafty man who always managed to obtain more than his share of everything.

His authority was based on the regard the other natives had for him as the Central Australian equivalent of a professor of theology. His memory was excellent, and like many other blind men, he was interested in poetry. When he was a young man he met Spencer, who witnessed his initiation in 1896. From the moment I met him he proceeded to take charge of our interviews and to arrange my dealings with the other native informants. He stated that he knew very well what I intended to do, and that he would manage everything for me.

One day, after he had dictated a myth about the moon, I asked him if he knew any myths about the sun. He replied that he did, but that it would not be worth my while to copy them because they could all be found in Spencer's book. He knew that Spencer had written a book because it was on my table and I referred to it frequently as we worked. I checked it and found that Yirramba was not mistaken. When I asked him how he could remember after so many years just which myths he had told Spencer,[3] he answered that the sun ceremony had been performed at his initiation (which took place before he lost his eyesight), and he remembered everything that took place at that momentous period of his life.

Tnyetika of Middleton Ponds was similar to Yirramba. I did not know him as well as I knew Yirramba, but he appeared to be the same sort of shrewd, intellectual old man. He knew many songs and myths and could always be counted upon to make the best of any situation. His authority was greater than that of Yirramba because he was living among his own people. He was an authority on traditional lore as well as the custodian of the most valued *tjurunga* in his region.

Moses, another blind man, was regarded as the chief of the Christian Aranda. His authority rested not, as usual, on the ownership of *tjurunga* and myths or on his knowledge of mythology, but upon his association with the white man. He assisted Strehlow in translating the Bible into Aranda and was the first native preacher of the new faith. Although he knew a great deal about the old faith as well as the new, once having imparted his knowledge to Strehlow he shared the zeal of that missionary and was loath to speak on the topic. He would never dictate a whole song or myth to me. He liked instead to be consulted on knotty problems of both the old and new religions. He was a dignified, kindly man, a real father imago.

Renana's position was somewhere between those of Yirramba and Moses. Although he too was a preacher of the

Christian faith, he still considered the *tjurunga* sacred and would not speak of them in the presence of the uninitiated or show them to a woman. Reverend Albrecht told me that whenever Renana stayed at the Mission for a few days, he would complain of a headache. As soon as he returned to Aroulbmolbma, the headache disappeared. At Hermannsburg he was merely a "Mission native," subject to the authority of the white man and without prestige; at Aroulbmolbma he was a chief in his own right. He and several young men had been sent to Aroulbmolbma some years before I met him, to care for the Mission goats. By chance he happened to be the hereditary chief of Aroulbmolbma, and could play there the role he so desperately wanted and, on a limited scale, live the life of his fathers. Renana once described to me a famous chief of ancient times. He was a man with a big face, long black hair, and a long beard. Hair also grew out of his nose and ears. He had only one eye and was completely covered with hair. Everything about him was huge: his face, his head, his belly. He was clever, bold, and famed far and wide for his skill with the spear and the knife. This description sounds like a child's anxiety distorted vision of the father in the role of a folk tale ogre.

Knatata, usually known as John, was the hereditary chief of Ndaria. He was an insignificant, surly man with little traditional knowledge. His chiefship was not taken seriously by anyone, although he probably would have had more influence if Ndaria had not been transformed into Hermannsburg. He would then have functioned as the organizer and leader of the ceremonies.

From the descriptions above, we can see that, since almost every grown man is a chief, and since the formally defined rights and prerogatives of the chief are few, a chiefship actually represented a potential source of authority for the aggressive and able man.

NOTES

1. S. D. Porteus, *The Psychology of a Primitive People* (New York: Longmans & Green, 1931), p. 264.
2. Carl Strehlow, *Die Aranda-und Loritjastämme* (Frankfurt: Baer, 1908), vol. IV, pt. I, pp. 42–43.
3. B. Spencer, *The Native Tribes of the Northern Territory of Australia*, 1914.

4

THE CHILDREN

Childbirth

According to Strehlow,[1] when an Aranda woman first feels the pangs of childbirth, she leaves her hut and goes to the women's camp. There her mother-in-law erects a separate hut, called the *waramba,* for her. She reclines in a semi-recumbent position, the upper half of her body somewhat elevated, while her hands press the earth behind her. The women of the camp assemble at the hut. They rub her body in the belief that they can separate the child from the mother in this way. Parturition occurs quickly and usually without difficulty. The umbilical cord is cut from the placenta with a sharp stone, and the placenta is buried. Hot ashes are placed on the navel of the child in order to dry it as quickly as possible. The infant is rubbed with eucalyptus bark and placed on a bed of soft bark. When the umbilical cord has dried, it is wound with string and placed on the baby's neck to promote his growth.

If a woman bears twins, the first born is called *aldoparinja*[2] and is believed to be the child of an evil wind which entered into the already pregnant woman. This child is a demon. The grandmother puts coal or sand into the newborn infant's mouth or beats its head with a stick. If the baby's body is covered with hair, or if it is in any other way malformed or unusual, it is killed in the same way, since it is regarded as the incarnation of an evil being.

The infants have rather pale red skins which become darker within a few weeks. The ceremony of smoking the child and the mother is performed in order to hasten this process and increase the milk supply of the mother. Some days after the birth of the child, the grandmother places her grandchild in a big bark trough and carries him to a suitable spot. There she digs a hole, fills it with green twigs, and erects a kind of platform above it. The mother and infant sit on the platform while the grandmother sets fire to the twigs. The smoke soon envelops the pair. The grandmother strokes the face, breasts, back, and abdomen of her daughter-in-law with twigs in order that she may have a plentiful supply of milk. The mother then squats over the fire in order to stop the flow of blood.

When they return to the hut in which the child was born, the mother-in-law cuts her daughter-in-law's hair and singes off the hair of the child. If the hair with which the child was born continued to grow, a demon might notice the child and eat it. With a piece of charcoal the grandmother draws circles on the child's face and body in order to make it ugly, since the demons eat only good and pretty children. She then presses the infant's nose into the shape which the Australian aborigines regard as pretty, and decorates the body of her daughter-in-law with red ochre.

In general, the information that I obtained on birth customs confirms Strehlow's data, although some details (e.g., the birth hut) appear to be variable and were not reported by my informants, while others appear to have been deliberately withheld from Strehlow, for reasons I will discuss later.

When a Yumu or Pindupi woman feels the first labor pains, she goes to the women's camp accompanied by two women who stand in the relationship of mother's brother's daughter to her. They rub her body and try to determine the position of the fetus. If the fetus is not in the right position, the women attempt to turn it around by manipulating the woman's abdomen. It is believed that if the child lies in a crosswise position, it might try to straighten its arms and cause the mother great pain. The umbilical cord is cut at some distance from the navel lest the child bleed to death. The maternal grandmother and a paternal aunt place warm ashes and sand on the mother's belly, and then they smoke her. The mother squats with her legs apart above the fire while the smoke goes into her vulva and stops the bleeding. Other explanations for the smoking were that it prevents the baby from drinking too much milk, which might make him too brown, or that it makes the breasts of the mother big. The umbilicus is wound with string and placed on the baby's neck in order to make him big and fat and prevent him from crying. It is believed that a woman's genitals have a foul odor after she has borne a child, so she must be smoked a second time before she can have intercourse with her husband.

Another custom, observed by all the tribes with which I came into contact, may be regarded as symbolic of *repression*. The grandmother, having heated her thumb in ashes, presses down on the penis of her grandchild to prevent it from growing too big. The clitoris of the infant girl is pressed back with the grandmother's heated heel for the same reason. At times the child's cheeks are struck with a centipede which is then placed on the child's genital. This, too, is done in an effort to keep the genitalia small. The tribes believe that, were these rites not practiced, the child might become the sort of man who gets an erection whenever he sees a woman.

The last few customs, although universally practiced, were not mentioned by Strehlow. They demonstrate the fact that human society or civilization is built on the repression of the libidinal impulses, that culture is evolved at the expense of gratification of the primal impulses. (It is also of some importance to note that the female ritual is not directed against the vagina, but against the clitoris.) From the point of view of my own ontogenetic theory of culture, these customs might be regarded as the specific traumata which determine the structure of the sublimations (i.e., of the culture) of the Central Australians. However, the experience is not actually traumatic for the infant, since it occurs immediately after birth and is often more a symbolic than an actual performance. The fact that the Australian men habitually disclaim possession of a big penis and of being libidinous should be regarded as a culturally determined parallel of this experience rather than as a consequence of it.

While the mother lies in the hut in which the child was born, the father may visit neither her nor his child. However, he continues to supply her with meat. If the child cries a great deal, it is believed that he is demanding the big *tjurunga*, which is then presumed to have been lost. It is therefore found, i.e., made, by the father, who winds it around with string and has it placed in the cradle under the infant's head.

After some weeks of separation from his wife, the father leaves the immediate area and goes to another camp. He returns decorated with red ochre. Having stayed in the women's camp for about a month, the woman returns to her husband with her baby and they resume their life together.

When the woman is no longer able to nurse her child, she and her husband abstain from coitus. They regard the inability to nurse as a sign that the woman is pregnant, and if they do not abstain the penis might go into the unborn child's eye and blind him, or make him ugly by pressing his nose out of shape. Almost all congenital malformations are believed to be due to parental indifference to this taboo.

Infants are carried about under the mother's arm, on the mother's hip, or in a bark cradle. The child is suckled for a long time, frequently for four or five years. Often an older child will push a younger one away from the breast and begin to suckle himself. The mother does not reject the elder child. A woman will rarely withhold her milk, even when the child is not her own. This was especially notable in the case of Tuma, who, although she had a one-year-old baby and was envious of the favors that Aldinga received from my wife, suckled Aldinga, a four year old, whenever he demanded the breast.

Naming the Child

When the child is about a year old and has begun to walk, he is given a name. The paternal grandfather generally decides on the name, which is usually that of the child's double, i.e., the ancestor who threw the bull-roarer at the mother and caused her to conceive. A child of the Wild Cat totem is usually called Tjilpa (Wild Cat), while a child of the Emu totem may be called Iliapa (Emu Feather). If the woman first became aware of her pregnancy while she was gathering figs, the child may be called Fig. The name may also be derived from some personal peculiarity or deformity.

Strehlow's statements on naming (as described above) are correct only with regard to the Aranda. The Lurittya group of tribes have a double naming system which is very confusing, since both types of name are employed simultaneously. Each individual has two names, a *tamu* name (given by the paternal grandfather) and a *tukutita* name (derived from the mythological ancestor).

Among the Pitjentara, the child is given his *tukutita* name when he is about one year old and his *tamu* name after he has been initiated. My informants told me that if it were done otherwise, the child might be eaten by demons. This statement clearly shows that the demons are projections of the immediate family, for, as we shall see, small children are in actual danger of being eaten by their parents. This danger ceases to exist after the child has received a name. While I was doing fieldwork in the west, Kanakana's child had no name. He was given a name when we returned to the Mission. Since he belonged to the Wild Cat totem, he was called Tjijanku (Little Pad) because the totemic ancestors wore little pads in their hair. The *tukutita* name of the boy whose initiation I witnessed[3] was Tapaltari, but after his initiation he was called Nyinga tukutita, which was his *tamu* name.

Among the Yumu and the Pindupi tribes, the *tamu* name is given first, and the *tukutita* name is given at the initiation ceremony. Thus, while Urantukutu was called Tapilkna as a child, he was called Urantukutu, his "sacred" name received at his initiation, as an adult. The Nambutji have the same system as the Aranda. Each individual has only one name, derived from the mythical ancestors.

Infanticide and Cannibalism

As soon as Tjijanku had a name, the great danger that threatens every child of the tribes west of the Aranda disappeared. He would not be eaten. The *tjurunga* protects the tribesmen from the cannibal demons; the name which links the child to the ancestral cult fulfills the same function with regard to his cannibal parents and brethren. In relating one of his dreams, Pukuti-wara mentioned a place called Kunanpiri (Bird's Excrement). He said that the bones of many children who had been eaten by their mothers were to be found there. He then remarked that the custom of eating children was spreading because of the great drought. The children were thin and the mothers hungry. Years ago it had been the custom for every second child to be eaten by the preceding child. This was exceedingly practical, especially when there were many children. Tjintjewara, an old Matuntara woman who lived at the Mission, was reported to have eaten her own sister. Ikitanpi, one of my informants from Tempe Downs, said that every second child was eaten in order to increase the strength and growth of the others. He remembered how Tjintjewara's mother killed her own child and gave it to Tjintjewara to eat. Tjintjewara admitted this much, but denied that she had actually eaten her sister. She said that when her father saw what had happened, he beat her and her mother and took the roasted baby away from her. In all probability, Tjintjewara did eat her sister; the details about her father were probably invented in an effort to accommodate herself to the ways of the white man.

The custom of eating children generally took two forms. When the Yumu, Pindupi, Ngali, or Nambutji were hungry, they ate small children with neither ceremonial nor animistic motives. Among the southern tribes, the Matuntara, Mularatara, or Pitjentara, every second child was eaten in the belief that the strength of the first child would be doubled by such a procedure. But even in these southern tribes the custom presents different aspects, depending on whether the information is obtained from a man or a woman. The men appear to practice infanticide and cannibalism because of moral principles, and the women because of hunger.

The Central Australians induce abortions for the purpose of eating the embryo by pressing on the abdomens of pregnant women. Patjili, a Ngali woman, told me that the Ngalis and the Yumus ate their own unborn children. The fetus was pulled out by the head, the placenta burnt, and the fetus roasted and eaten by the mother and the children. The mother killed the fetus because she was hungry for meat, while the children ate the fetus because they were told that it would make them big and strong. Iwana, a Matuntara woman, told me that a little baby would be killed by the mother, while a big boy like Jankitji would be killed by the father by being beaten on the head. The father would not eat the child, but would give it to the mother, who would share it among the other children. Iwana said that even a child as big as Aldinga might be eaten.

Jankitji, Uluru, and Aldinga had each eaten one of their brothers. When I questioned Aldinga on the subject, it was obvious that he did not like to talk about it. He admitted only part of his role in the situation, and said that Jankitji and Uluru had eaten the other brothers while he had been merely an onlooker. Iwana admitted having killed and eaten her small daughter but said that her husband had cut her on the head, breast, and arms for doing so. I was not able to tell whether her husband's action was customary in such cases, or whether it was his individual reaction.

While, according to most informants, children are not eaten after they have received a name, Iwana stated that even Aldinga and Jankitji might be eaten. Probably the former testimony concerns the "ceremonial" aspect, and the latter is the "practical" or famine aspect of the custom. It is interesting to note that the custom was evidently regarded as not quite "right," even by those who practiced it. The men abstained from the meat, and at times even punished the women for what they had done.

We can characterize Central Australian culture as one in which fantasies come closer to realization, and the borderline between thought and deed is narrower than in other human societies. Instead of the fantasy of the cannibal parents or brothers, in Australia there are real cannibals in the immediate family. How can we explain the fact that the mother who is generally so good and nonresisting in the treatment of her children is also capable of killing them? The answer evidently lies in the "unorganized" character of their primitive psyche. Opposite and contradictory emotions and trends do not balance each other and result in compromise formations. Instead, each trend is permitted complete sway at the moment of its ascendancy. For the most part, however, the parents continue to be protective, projecting their cannibalistic and libidinal propensities onto the demons and thus enabling society to continue.

Disciplining the Child

The discipline of children is a topic that occupies parents throughout our own culture. It is often said that primitive peoples, particularly the Central Australians, never punish their children. Reverend Albrecht related some of his experiences on this matter. He had punished Rufus, a boy of about eight, by locking him in the boys' house. The child roared and screamed for about an hour, at which point the Missionary gave him a beating. When the natives who lived at the station saw this, they threatened to kill the Reverend because he had been beating a child—an unheard-of thing. When they became somewhat calmer, Moses and some of the other elders visited Reverend Albrecht to talk matters over. The Reverend explained that it was wrong to allow their children to have their own way at all times, that such treatment resulted in the children's throwing stones at their fathers and mothers. The old men declared that he was right, and the next day, when the Reverend wished to pursue the topic further, the old men ran away because they were ashamed of their former conduct.

The difference between native and Western pedagogical ideals is apparent in the affair of the boy who was a monitor for the school children. He beat the little children with a whip and was nearly killed by the natives for doing so. They said that such things might be done by a white or by a half-caste, but never by one of their own race. It would be an exaggeration to say that the natives never beat or punish their children or thwart their children's desires. When a child becomes too trying, an adult may lose his patience and give him a slap. This is not exactly a punishment, in the same sense as the punishment that occurred when Lelil-tukutu was beaten on the behind for having unwittingly sat on an ancestral stone.

Scoldings are used far more frequently than corporal measures. For instance, one evening when the children had been asked to make less noise and had refused, Ankili, a Pitjentara woman, shouted at them: "*Kuna tjikintaku!*" ("Drink excrement or vagina!") When Urantukutu threw stones at his father, the old man would shout: "*Kuna tjikintaku runkani?*" ("Drink excrement! How dare you throw a stone?")

Urantukutu told me that he had been a very pugnacious little boy who was always throwing stones at adults. His narrative typifies the general native attitude toward children and gives a picture of the Australian pedagogical methods. When he was very small, he stayed constantly in the camp with his mother. When he grew a little older his mother would often say to him: "You see that willy wagtail? It is scolding you. Go and kill it!" He would then throw a stick at it. Or she would say: "There is a red breast. It is swearing. Go and kill it!" Since she did not want him to wander far from the camp, whenever he did so she would shout: "Do you see that black stump? It is an enemy, a blood avenger!" He would run back to his mother and howl: "Mother, come and pick me up," and she would lift him and carry him on her shoulder.

What is it that the Australians resent so violently in Western methods of corporal punishment, since they themselves are quite capable of slapping or beating a child? We can approach this question by comparing the behavior of the children at the Mission with the behavior of the desert children. At the Mission, Depitarinja and some of the others might frequently be seen playing the "sadistic" game. Depitarinja would give orders to a smaller boy or girl while the victim stood at attention and received any slaps that his tormentor might wish to administer. Such games are unknown in the bush. While a bigger boy will often use his strength to coerce a smaller child, he will never make a deliberate use of power and take pleasure in the submission of others. Obviously the children at the Mission were influenced by the white school methods, and it is the deliberate infliction of pain, the *sadistic pedagogy of the white man,* which so scandalized the natives that they attacked the Missionary, Reverend Albrecht.

Central Australians, like all other parents, indulge to a considerable extent in frightening their children. They tell their childen: "Don't go far away. A *bankalanga* might come. He will take you to his cave and cook and eat you." Or they will say: "The Nyipandipandi (Wild-woman) will come. She will put you in her trough and take you away. Then you will be her child and she will make you like herself. You will never see your own father and mother again."

The parents were certainly being realistic in their desire for their children to stay near the camp, because children are easily lost in the bush. There is also the possibility that they will be killed by strangers. However, we can look at the threats of the parents from another point of view. The parents were actually threatening the children with themselves as "bad," in other words, as sexual objects (e.g., "The Nyipandipandi will . . . take you away. You will be her child. . . ."). The anxiety dreams related by the children during the play sessions revealed the mother as an earthly prototype of the phallic demon.

A three-year-old boy dreamed that a huge kangaroo was trying to scratch out his eyes. The child tried to push the paws away, but in vain. Obviously the anxiety was related to something that the child should not have seen, probably a coitus scene. The same child was also dreadfully frightened of the emu demon, one of the favorite bogeys of the nursery.

We may therefore say that if we consider the four main techniques of pedagogy—erection of an ego-ideal, use of a bogey to create fear, corporal punishment, and preaching—the first two predominate, while the third and fourth are nearly absent in the Central Australian nursery. On the whole, it can be said that the parents rule with a lenient hand, and that techniques of intimidation are brought into action only at the threshold of adult life.

Childhood and the Game

Since I, as a psychoanalyst, am more interested in inner meaning than in appearance, it is not necessary to repeat observations made by others on the play and life of the Central Australian children. We will attempt to discover how universal conflicts are expressed in the play of the Australian children, and also how the particular characteristics of Australian society and culture mold both the conflicts and their externalization. We will be concerned mainly with description and analysis of the events which occurred during forty-four play sessions which I conducted at the Mission station and in the bush.[4]

Children the world over play, in actuality, only one game—that of being grown up.[5] In almost all areas of Australian life the unconscious fantasies of the people shine through their behavior with a clarity that is almost startling. Children are always more transparent than adults, and the Central Australian children present us with an unequaled view of the unconscious life of childhood.

In general, the Australian children play only two games, of which there are many variations. One game is called *wati-wati* (men-men) by the boys and *minma-minma* (women-women) by the girls of the Pitjentara and Mularatara tribes. Before beginning the game the children erect a windbreak, creating in this way a nursery, an artificial barrier between themselves and reality. The boys attack the mounds of the termite ants with their miniature spears and present the ants to the girls with the same pride shown by full-grown men when they present their women with kangaroos they have just killed. The game of the girls consists of cooking these "kangaroos" and nursing the twigs which they call their babies. The sand of the desert becomes the seeds and roots gathered and prepared for the "husbands" and "children."

The other game of the Central Australian children is the oracle game, which the Aranda call *altjira* (ancestor). Eucalyptus leaves represent a married couple and child; other leaves become the women's camp and the boys' camp. The children tap the leaves, or, by hitting them with twigs, make them fall on top of one another. The way the leaves fall is interpreted as a prediction of the future life of the player as a husband, wife, or simply an adult member of the tribe. The Pitjentara and Mularatara have several names for this game: *nalpi-punganyi* (hit leaves), *milpa-tunanyi* (use the crooked stick), and *tjukurpa*, which corresponds to the Aranda *altjira.*

In the narratives of many of the women who acted as my informants, such as Aliumba, a Ngatatara, and Ilpaltalaka, an Aranda, the two games, "playing at being grown up" and the "oracle game," were continually being confused. Aliumba gave the following description of the game of "women-women." The girls made a little hut. They collected the leaves of the gum tree and looked at the imprints made by the leaves in the sand. They made dolls of the branches of the punti bush and carried them about in bark troughs, as their mothers did. They imitated the way that women fought, first folding their hands and then hitting each other on the head with yam sticks. They struck the *kwata-kililia* (a plant which is the totemic representative of girls) and reviled it with such expletives as "stinky vagina." They shouted: "You have taken our husbands away, and now they give you all the meat." They treated the plant in every way as though it were a grown woman, a sister, mother, or grandmother, and acted as though they were themselves grown women involved in a jealous fight. When the fight with the plant paled, they continued the argument with one another. The girls made houses of gum leaves and then destroyed them by hitting the edifices with their crooked sticks. They prophesied the future from the imprints the fallen leaves left on the sand.

Although this is a fairly complete account of the description of the play of the children, it is not possible to tell whether it was the children themselves who confused the two games, or whether the confusion occurred later in the memories of the informants.

Urantukutu described the game of the boys. They pretended that the white ants were kangaroos, killed them, and with great pride carried them back to the place where the girls were playing. Occasionally one boy would hold a piece of rolled bark while the other boys speared it. The boys fought among themselves, spearing one another in the leg in imitation of the men. In this game they used their fingers as though they were knives. They also speared one another with long pieces of grass, defending themselves with miniature shields, and learned the art of fighting in the process. Urantukutu's first memory is of receiving a little spear which he straightened in a camp fire.

Kanakana, a Mularatara, added a detail of importance to our knowledge of the oracle game. She said that when the fig leaves were set up in a row, one of them represented a man and another a woman. Saliva, which represented both vaginal discharge and semen, was placed on the leaves, which were then rubbed together. "Now," the children would say, "they are cohabiting."

Although the name of the "playing at being grown up" game differs in the different tribes, it can always be translated as "men-men" or "women-women," which helps us in discovering the meaning of the game from the name alone. The oracle game, whether called *altjira* or *tukurpa,* is translated as "ancestor," a name which at first tells us little about the nature of the game. However, we know from our research that these words originally meant "dream," so the parallel between the wish-fulfilling aspects of the prophecy game and the dream becomes apparent. Children either play (i.e., act out, in the fullest sense of the term) or dream their desires. Moreover, we can now understand the frequent lapses of the informants in confusing the oracle game with the "men" or "women" game, for the source of much unrest, the origin of the ungratified wishes of the child, is the fact that he is a child and not an adult. To the child, the adult seems capable of gratifying all those instinctual desires that his smallness and dependent position prevent him from doing. The adult, unknown to the child, is also unable to gratify all of his instinctual desires, and so he too plays these games, calling them not play, but ritual. The adults play *altjira;* the legendary heroes are *altjira.* When the adults make marks in the sand during the ritual of "remaking the footsteps of the ancestors," they are really children pretending that they are grown up. They play the same games as the children, but under the patronage of the ancestors (i.e., of the superego).

Play Sessions at the Mission

I played *altjira* with the children in a way which permitted me to gain some insight into the problems of childhood and growing up. I proceeded on the scheme of Melanie Klein's play-analysis,[6] but omitted the interpretations. I spent twenty-three hours during a period of about two months with four children at Hermannsburg. One of the children, Depitarinja, a ten-year-old boy, was present at all the meetings. Nyiki, a boy of about eight, was present at all but one. The two little girls, Tena and Angelica, who were five and seven years old, were absent from two meetings each. A few other children came to our play sessions once or twice, but the only one who did so with any regularity was the half-Aranda, half-Afghan Nomi, a three-year-old boy who was present at five consecutive meetings.

In order to supplement the makeshift playthings which the desert provided—sand, leaves, twigs, and so on—I supplied the children with the following toys: the figure of a kind of ape holding a tree,[7] a small figure of a goat, a rubber snake, a paper trumpet, a mirror, a large India rubber doll, a transparent water-filled globe in which floated an artificial fish and stars, a water pistol, a rubber head with a protruding snake-like tongue, several illustrated magazines, and paper and pencils. The children were introduced to the toys gradually.

Hour 1

The first play hour[8] was spent in overcoming the natural shyness of the children. Their question seemed to be: "What does he want?" To them, the white man was an object of awe and alarm. Although they accepted the candies I offered them, they remained puzzled and unresponsive.

Hour 2

On their second visit, two of the children at least understood that they were to do and say whatever they wished.

> Nyiki told the following story: he left the camp, found a lizard, killed it, cut off its head, and ate the rest. He returned to the camp, but soon left again. This time he killed an euro.

I expressed astonishment that he should have killed an euro.

> Nyiki replied that he had a little spear and that he killed the animal with that. He said that when he returned to the camp the second time, he went for a swim in the Finke River.

This narrative interests us because it illustrates the transition from fact to fantasy. Nyiki was too small to have killed an euro, although he had probably killed many lizards. The Finke River had not had enough water for swimming since before he was born. The narratives of his elders supplied the content as well as the form for his narrative.

> He spoke of other subjects: the Mission teacher, the Missionary, etc.

I then asked him to tell me some of his recent dreams.

> He spoke of two. In one, an *erintja* (demon) had frightened him, and in the second a *ltana* (ghost) had done the same thing. Both dreams consisted of a single image. There was no action.

I gave him paper and pencil. The picture he drew contained the clue to the meaning of the anxiety in the dream.

> He drew a demon with an erection. Next to the demon was a *tana*, the wooden trough in which the women carry their children. He said that his mother had told him that if he was bad, the *ltana* would carry him off in a *tana*.

I asked him whether he had been bad.

> He replied that he had been, and had said "*para takia*" (erected penis) to his mother before he went to sleep.

The demon with the erect penis is obviously the "bad" mother.

> As he was speaking, he played with the scabs on his legs. He then drew a snake in the sand. He began to speak of ghosts in general. He said that his father had recently died.

Hour 3

The children were again unresponsive. I brought out the monkey, the goat, the rubber snake, the paper trumpet, and the mirror.

> The children's attention was riveted on the monkey. Depitarinja decided that it had hanging testicles, then a vagina, and finally, a rectum. He elaborated his description. The monkey had a hairy vagina, a hairy rectum; it had breasts. He finished by declaring that the monkey was a woman.

It is interesting to note that, although the monkey lacked sexual characteristics, the first thing that Depitarinja remarked was that it had testicles, and hanging ones at that. However, he finally decided that it was a woman.

> The children then played with the goat, which they called a little girl. They made it *aruntjima* (kiss, lick) the monkey, first on the mouth and then on the vagina. They brought the three animals together. They now treated the goat as a male. It cohabited with the monkey, while the snake had intercourse with it from behind.[9]

The goat was a male in relation to the monkey and a female in relation to the snake.

> The snake then attracted the attention of the group. Depitarinja and Nyiki poked it at the girls, who behaved in a

> manner typical of girls everywhere; they were ashamed. They giggled and turned away.

This behavior seems inappropriate if we imagine that the girls regarded the rubber snake as a real one. But if we consider the phallic significance of the serpent, we at once understand their behavior.

> The girls made a few half-hearted attempts to grab the snake, but they did not dare to carry through their intentions. Depitarinja then made the snake cohabit with the monkey and suck its nipples. The children decided that the monkey was a woman; the goat, a little girl; and the snake, a boy.

This distribution of roles was the result of the boys' desires in the family situation; i.e., there was no father. Depitarinja fulfilled his Oedipal desires in the game. The snake had intercourse with the same woman from whom he obtained milk. This woman in Depitarinja's life was not his mother, but his grandmother.

> Depitarinja said that he too dreamt of *erintjas* frequently. He asked me for pencil and paper and drew one of his dream demons.

He first dreamt of a demon when he had stolen something from his grandmother. Since that time he had dreamed of demons often.

> The little girls became very interested in the mirror, as did the boys to a lesser extent. They giggled in an embarrassed way as they looked at their images. The little group became more and more excited. Finally, they put all the toys together, forming a mountain in which each toy was made to cohabit with the one nearest it. *They looked at the image of this mass cohabitation in the mirror.*

Hour 4

Nyiki and one of the little girls were absent from this meeting. Depitarinja had by this time taken charge of the proceedings. He brought another little boy and girl to take the place of the absentees.

> The children sat and played in the sand. Their game was in the mythological style.

This is significant in that myths are taboo for children, and they most probably had never heard a single narrative of this kind. However, that style of speaking so permeated the culture that its use by the children was inevitable.

> They made marks in the sand. "This is a soakage. Now an euro comes to drink. There is a hut. Now the goat wants to drink water." The snake bit the woman who was represented by the monkey. A camp was represented by concentric circles just as it is on the *tjurunga*.

The game was extraordinarily similar to the *altjiranga* (ancestor) stories of the adults, both in the geographical representation of everyday events and, in places, the similarity of the symbols used.

> The children turned their attention to the toys once more. One of the little girls declared that the snake was her favorite toy. She made it cohabit with the monkey in the normal position and *a tergo*. She then looked at me in a coquettish and provocative manner. She pretended that she was afraid of the snake and asked Depitarinja to protect her. She decided that the snake was a boy and put it into the paper trumpet and then pulled it out again.

She was evidently acting out the sequel of the snake's cohabitation, i.e., pregnancy and parturition. Her symbolism was also used by the adults in their religious rites. The child-snake-ancestor procreates itself and then emerges from the rock (which is what the children called the paper trumpet), which is the symbol of the mother.

> The other little girl told me that she dreamt of an *erintja*, who was all bones.

She slept beside her mother, who was very ill and quite thin.

Hour 5

The children played their own games. The boys pretended to spear one another while the girls hit each other with sticks. The girls built a small hut and pretended that they were mothers. Their old style of play merged with the new; the goat was their child. They drew on the sand. Concentric circles again indicated the camp, and water was represented by a number of parallel lines.

Hour 6

Depitarinja was hostile to me. He drew a portrait, which he claimed was of Nyiki, but which was actually a caricature of me. Angelica said that she had dreamt of someone with a very big nose. She decided that it was her classificatory mother, a woman who had refused to give her some food before she went to sleep the night before. Then she decided that it was an *erintja* who looked like her classificatory mother.

The mother, by refusing to grant the child's wishes, was transformed from something essentially good (a human being) into something essentially bad (a demon). When the negative feelings toward the mother are transferred to the demon representative, the positive feelings toward her are preserved. The demon also represented the phallic mother (note the big nose), and the phallus is the source of pleasure and of danger from the infantile point of view.

Hour 7

I played "soccer" with the children. They all enjoyed it and became very lively. Depitarinja was talkative. He related several long narratives about killing and eating snakes and lizards. He spoke about playing with bullocks which he represented with stones. He mentioned the well which had been dug at the Mission before Christmas. Tena had fallen into the well. The other children told of the races arranged

> by the Mission at Christmas time. Depitarinja had won a doll which he kissed whenever he felt himself to be in love with a girl. As he talked, he made the snake bite the rectum of the ape.

Hour 8

Tena was absent from this meeting. Another little girl took her place.

> The children said that Tena had not come because she was "ashamed" because we had spoken of her having fallen into the well. The new child drew a ghost which had a huge protruding stomach and no penis. She stated that the ghost ate grass, and that it had bitten a boy's toe the day before. She then drew a demon, on which the penis was very conspicuous.

Prompted by my own curiosity, I asked her why the demon had a penis when the ghost did not. I should not have asked this question, for in doing so I influenced her.

> The child solved the problem quite simply by supplying the ghost with a penis. The girls looked at the penis of the demon without any of the embarrassment which they had felt in the case of the rubber snake. Depitarinja seemed to feel that the attention of the children was riveted too exclusively on the penis of the demon, and pointed to its anus.

Hour 9

After an interruption of about two weeks, we resumed our sessions with all the children present. The toys I gave them to play with were the monkey, the snake, a girl-in-the-box, a jack-in-the-box, and a new paper trumpet.

During the two weeks when we had had no meetings, Depitarinja had gotten into considerable trouble at the Mission because of his stealing. He continually stole food from the women in the camp and from the cook of the white teachers, who was his classificatory mother. He had also had a fight with another boy who was beating his classificatory sister.

> The children were playing with the toys, placing them next to one another and declaring that the toys were talking.

I made a remark to Nyiki about Depitarinja's stealing and he took the hint.

> Nyiki used my remark as the starting point of a game. The monkey (a woman) stole the girl from the box (hut) and tried to carry her off into a stone hut like a demon. Then the snake tried to bite and steal the child. Depitarinja joined in the game. He made the snake cohabit with the monkey and the child. A new variation appeared in the theme of the game. The snake was made to bite the monkey (woman) because she had tried to steal the child. The snake bit the breast of the monkey and licked and kissed both her anus and vagina. Depitarinja declared that the monkey was going to marry the girl.

I remarked that both the monkey and the girl were women, and asked Depitarinja how they could marry each other.

> "Well," he said, "then they will be friends." He placed the monkey on top of the snake and declared that they were cohabiting.

I objected to this on the grounds that the snake had always represented the male and the monkey the female, and therefore their positions should have been reversed.

> He agreed with me, but stated that the monkey was pregnant, and because of that was permitted to lie on top of the snake. After a moment, he added that the girl was pregnant too.

What information does this game of Depitarinja's give us in regard to his stealing? For one thing, we see that stealing, cohabiting, biting, and attacking were regarded as parallel activities. There is also linguistic evidence for this assumption.

The word *nyilkna* means both theft and illicit intercourse. We can interpret the action of the monkey and the snake in stealing the child as another way of saying that they were having intercourse, that they were making a child.

Depitarinja was constantly trying to steal things from the women. The really curious feature in his game was the marriage of the monkey (woman) to the girl, which I think is an example of the phallic mother. What Depitarinja constantly tried to steal from the women was the phallus. The fact that he decided that the monkey was pregnant adds weight to this interpretation, since a pregnant woman is one who cohabits in the male position. The child is her phallus.

His positive Oedipal trends were acted out by the snake who had intercourse with the woman (mother) and the child (sister) in quick succession. As mentioned above, he had just had a fight with a boy who had beaten his classificatory sister.

Hour 10

> Depitarinja continued the theme of the previous session. The mother was going to marry the girl-in-the-box, her daughter.

I asked him how that could happen.

> He then decided that the snake would marry the girl. The snake married three women: a mother and her two daughters. The story then changed. The snake married only the mother. The two daughters were his own. Nyiki remarked that he had again dreamt of demons. He drew pictures of them, one big one and one small one. Neither of them had the usual huge penis. Their feet were malformed instead.

Hour 11

The two little girls who usually came were not present. In their place came two Christian Aranda girls, Gwenda and

Irma. I gave them the usual toys and introduced them to a new one, the water-filled celluloid globe.

> The new toy attracted a different kind of interest than had the other toys. The object was of interest because of its real, not symbolic, significance. They called the star a crayfish; the alligator, a lizard; and the fish was recognized as a fish. They shook the globe and made the figures move about. Irma then played with the goat and Gwenda with the snake. At one time or another, during the afternoon, each of the children tried to penetrate into the center of the globe with the snake. They repeated the same game with the monkey. Then, as if an explanation were necessary, Nyiki stated that *the snake wanted to go into the inside of its grandmother (the globe), in order to kill her.* The monkey was then used to represent the grandmother. She carried a wooden vessel represented by a piece of cork, in which there was food, represented by sand. They put the cork behind the monkey and declared that she was defecating into it.

It is interesting to note that the globe, which was later used to represent the mother's body, was first compared by the children to a witchetty grub, a great delicacy. The only point of resemblance between the two was that they were both striped. I also gave them another new toy, a grotesque rubber head with a protruding snakelike tongue.

> They said that the tongue was like an automobile.[10] One of the little girls said that she had dreamt of an automobile. The snakelike tongue and the rubber snake were placed together by Depitarinja, who then arranged the toys in a circle with the goat, which represented a boy, in the center. By blowing into a hole in the back of the rubber head, Depitarinja made the tongue pop out of the mouth. He promptly directed his attack against the girls, who retreated in great haste and confusion. He glanced at the toys and remarked that the goat was "teaching" the monkey.

Since the goat represented Depitarinja and the monkey represented an old woman, and the statement of their relationship followed right after his hostility to the girls, we may assume that he was acting out his hostility toward the mother.

> Depitarinja stated that the snakelike tongue was a ghost, and began a drama in which the ghost had come to carry off the mother.

These bogeys of infancy, the ghosts, are evidently the anxiety-laden representatives of primary infantile desires. The significance of these desires was revealed in the next scene of Depitarinja's drama.

> The snake, goat, globe, and tongue were placed at the anus of the monkey. He explained that they all wanted to kiss (*aruntjima*) but not to smell or lick (*injainama*) the monkey.

He made this distinction because of the influence of the Mission school. *Aruntijima* can really be translated, and is thought of by the Europeans of the area, as tenderness, while *injainama* is regarded by the Missionary as a perversion, although the terms are used almost interchangeably by the natives. When they told me that the other toys were kissing the monkey, I pantomimed the motions of coitus, and told the children that I thought they meant that, rather than kissing.

> They responded by laughing with great pleasure.

HOUR 12

Nyiki was absent and his place was taken by another boy, Konrad.

> Depitarinja told me that he had had a dream in which a rabbit with a big mouth chased him. He said that the rabbit had looked like Konrad.

Konrad had taken a lizard from Depitarinja a few days before this, and had also beaten him.

> Depitarinja then "shot"[11] the girl doll whom he called Francisca. Francisca was our Aranda servant girl. He added that

he was glad that he had shot the doll since she had refused him food.

Our servant girl had never refused him anything, but the day before my wife had denied his demand for bread and butter because she was busy. The "shooting" or sadistic coitus was therefore directed against my wife, as a mother substitute. After this meeting my work with the children was interrupted for a few weeks, because a new informant had arrived at Hermannsburg.

Hour 13

Nyiki did not appear at the next meeting. A new girl was present.

> The children were all a bit shy at the beginning of the session. Depitarinja began a narrative about a carpet snake. He had found it under a rock, pulled it out, killed it, and eaten it. Then he suddenly passed from fact to fiction. He said that Aldinga (another little boy) had told him about flying snakes, who lived in the tops of hollow trees and flew about in the branches. He suddenly jerked the snake out of the monkey's vagina, an act which caused all the other children to laugh.

Hour 14

> The children were all in high spirits. They brought their toy spears and pretended to spear one another. The little girls were very active and played in the same way as the boys during the spearing game. The girls shoved the snake into the monkey's vagina, just as Depitarinja had done the day before. Then Depitarinja took the lead in the play. He shoved both the snake and the goat into the monkey's vagina amidst general laughter. He said that they were licking and smelling the monkey.

I asked him whether he did that to the girls.

> He replied that he did and that it was very good. He said that he liked to lick the "wound," which is the children's word for vagina. He then slid his hand down the body of the snake and said that it had no penis.

This assertion is significant in that the snake was the toy which, until then, had openly been used as a phallic symbol. His remark gains in significance when it is noted that he made the statement just after having used the word "wound" for vagina, and after having admitted to having performed cunnilingus.

> Once again, he shoved the snake and the goat into the monkey's vagina, while declaring that they were going to pull the vagina out and eat it.

The word he used for "eat" (*ilkuma*) is often used euphemistically for coitus. This is an example of the idea of body-destruction which is implied in cohabitation. Another feature of interest in his behavior is the alternation of castration anxiety and primitive phallic-destructive trends.

> Both the boys and the girls enjoyed the idea of the snake and the goat eating the monkey's vagina.

I had given the children a water pistol. A basin of water was brought out for their use.

> Depitarinja filled the pistol with water and placed it over his penis. He squirted water at the monkey's nose and eyes. He then threw a ball with great accuracy at the vagina of one of the little girls. She threw the ball back at his penis with equal accuracy.

Hour 15

Depitarinja identified himself with the girl doll. He held the water pistol to his penis and called the water that he squirted out, "milk."

Hour 16

Depitarinja, who by now had appointed himself master of ceremonies, brought a new little boy, Edwin, to the session.

> In a decidedly sadistic manner, using great force, Depitarinja squirted water into the ears, eyes, and noses of the little girls. He then had a squirting competition with Edwin. When he fought with other boys, Depitarinja used his penis as a weapon. He kept them at a distance by urinating at them.

I then showed the children a Hungarian theatrical magazine.

> When either of the boys saw a picture of an actress, he immediately exclaimed that it was the wife of the other. Depitarinja identified himself with an elderly man who was standing with several girls on one of the pictures. Depitarinja declared that the girls were his own wives. He also identified himself with a picture of an actress in slacks. The other children laughed at him when I told them that it was a picture of a woman, not a boy.
>
> Edwin had dreamt of a demon with a red mouth. When I asked him to draw a picture of it, Depitarinja told him to draw the demon with an erect penis at the moment when he cohabits with his wife.

Depitarinja had by this time become permanently attached to our household. He had all his meals in our kitchen and returned to the boys' house only to sleep.

Hour 17

I had become the representative of the father-imago for Depitarinja. At this meeting Nomi, the three-year-old Aranda-Afghan child, was also present.

> Depitarinja spent most of this meeting acting out his fantasies of omnipotence and hostility against me. He shot everyone present, including myself, with the water pistol.

That this squirting with the water pistol was symbolic of urinating was shown by his remark:

> "This is how Nomi urinates." He squirted water at me and my dog. He dug a hole in the sand and said, "This is where Nomi urinates." He then squirted out the water very slowly and said, "This is the way that girls urinate."

Hour 18

The next meeting took place two days after the previous one. Depitarinja had gotten into serious trouble in the interim. He had been caught licking and smelling the genitalia of a four-year-old white girl. Her father had caught him and given him a terrible thrashing.

> Depitarinja looked very sad when he came to play.

I asked him what the trouble was.

> He told me that one woman had stolen the food of another woman and when the white man caught her, he had given her a beating. After having gone through this story, he finally told me the truth.

The lie that he told as an explanation for his depression shows the sexual meaning of stealing. That he identified himself with the snake was shown by the next game he played. In this game, the abreacting function of all the children's games was also apparent.

> He picked up the snake and declared that it was sad. Then he made it smell the vagina of the goat, after which he declared that the two were getting married. The monkey, who was usually the mother or the wife of the snake, became the snake's mother-in-law. She was therefore taboo for the snake. The mother-in-law chased the snake with a stick, and forced him to leave his food and run away. Then she ate his food. The big doll was called Jalpa, which was the name of one of the grown girls at the Mission. The other toys were boys who were talking to her through the window of her house. The monkey went home to her husband, a toy fish, who smelled her vagina and anus, which gave her great pleasure.

> Depitarinja shot the fish (father) with the water pistol. Then a chief came and beat all the toys for having smelled and licked the vagina and anus of the monkey.

This game was obviously a repetition and elaboration of the events of the day before. In the game, however, a multitude replaced the individual, and so lightened the burden of his guilt.

> The big rubber doll was first the child of the snake and the goat and then the mother of all the toys. She made all the toys smell and lick her vagina and anus.

Here again Depitarinja lightened his guilt. He was saying, in effect, that the little girl had initiated the sex play. It is also important to note that the doll was first a child (as was the little girl in reality), but then became the mother, as she must have been in Depitarinja's fantasy.

Hour 19

Nomi was present at this session.

> Depitarinja had come out of the depression of the last meeting and had entered a manic phase. He denied that there was anything troubling him. He played a game in which the monkey was Nyiki's wife, and the fish his brother-in-law. He said that Nyiki licked and smelled the vagina and anus of the monkey and the goat. He told Nyiki to eat the anus. In reply, Nyiki told him to eat the vulva. Depitarinja then declared that Nyiki put his penis into the eyes and nostrils of his wife. He reeled off a string of Nyiki's wives: the monkey, the goat, a rabbit, etc. He became very excited and, picking up the monkey, exclaimed that she was his own wife. During this whole procedure, he was hugging and kissing Nyiki and the two girls. He then used the water pistol to shoot everyone present, including myself. While doing this he said that he was Nomi, urinating. Then Nomi said that Nyiki was urinating. Depitarinja, not to be outdone, said that when Nyiki cohabited with his wife, his penis grew

> very big. Then it split, and then Nyiki split in half. Then Depitarinja dug a hole in the ground and said that he was "shooting."

This time "shooting" was definitely used in its colloquial sense, as a synonym for cohabiting.

> Nyiki, he declared, cohabited with the girls per se. He then declared that he and Nyiki were two demons.

His high spirits were due in part to the fact that my wife had just given him a new handkerchief.

> He looked at his reflection in the mirror in my room and kissed it several times.

Hour 20

> Nomi was present at this meeting. Nomi did exactly the same things with the ape that Depitarinja had done at the second meeting.

Nomi could not have done these things in imitation of Depitarinja, since he had not been a member of the group at that time.

> First he shoved the snake into the monkey's vagina, then he placed the snake at the monkey's breast and then at its mouth. Nomi and Depitarinja blew into each other's mouths through the paper trumpet and the water pistol. Then they began once more the game of "everybody cohabits with everybody else."
>
> They saw a picture in an illustrated magazine of two girls in bathing suits, standing side by side. Depitarinja declared that they were cohabiting.

I explained they could not do that, since they were both girls.

> Depitarinja agreed with me. He said that they could not cohabit with each other because neither of them had a penis.

> Depitarinja accused Nyiki of having had intercourse with a lizard, and was in turn accused of having cohabited with a mouse. Gradually, the accusations became more plausible. Depitarinja finally declared that Nyiki and another boy had put their arms into a camel's vagina up to the elbow. Nomi bore witness to the occurrence. Nyiki then related a tale about Depitarinja. A few days earlier, Depitarinja had caught hold of a smaller boy. He had pushed back the foreskin, rubbed sand on the penis, and masturbated the child until he ejaculated. Then he had had anal intercourse with the boy. Shortly after that, another boy in the boys' house had had anal intercourse with Depitarinja, and when he had finished, Depitarinja had done the same with him. Depitarinja was somewhat embarrassed while Nyiki told this story. He made several half-hearted denials, but they were made without real conviction. He tried to prove by his subsequent actions that he had nothing to do with that sort of thing. He took the water pistol and threatened the girls with it. They bowed their heads in humble submission and in expectation of the "shooting." Then he shot into the monkey's vagina and rectum with the pistol. He was most emphatically a man.
>
> Nyiki went on with his narrative. He accused Depitarinja of having tasted the feces of a camel. Depitarinja retorted that Nyiki had masturbated a mare in the same way that he himself had been accused of masturbating a camel.

Since there were no tame horses at the Mission station, Depitarinja's accusation was manifestly untrue, while it was quite probable that he had tasted the feces of the camel as Nyiki affirmed.

Hour 21

The same five children were present at this meeting as at the previous one. I gave them a rubber ball to play with.

> The children all got very excited as they threw the ball at each other. In the midst of this excitement, Nomi let go a *flatus*, to the general amusement of the others. He laughed with the rest and was not the least embarrassed.

Mrs. Róheim had told me that Depitarinja had become terribly frightened when he saw her cutting the neck of a chicken. I asked him what had frightened him so.

> He told me that the blood had upset him. When he was a very small child he had seen his older brother killing a bullock. That was the first time that he had seen blood. Turning to Nomi, he called him a little bullock. The children began once again to throw the ball and shoot one another with the pistol. Nomi, the smallest, got rather more than his share of having the ball thrown at him, and became sulky.
>
> Then Depitarinja began to talk about the bullock's blood again. While he was speaking of this, he shot himself in the mouth with the water pistol. He then declared that the monkey was Nyiki's wife, and, using the snake to represent Nyiki, showed how he licked his wife's vagina. He saw a picture in a magazine of a cow dressed like a woman. He called it a demon.

Hour 22

During this session I tried to get some material on screen memories. I asked Depitarinja and Nyiki to tell me their earliest memories.

> My question caused them both to become visibly depressed. After some thought, they both had something to tell me. Nyiki remembered seeing a kangaroo and exclaiming, "Meat." Depitarinja's first memory was of his mother giving him goat's milk to drink. He said that he had had a little knife at the time with which he cut his sister. His father had chased him with a stick and beaten him for this.

The pantomime which followed helps us to understand the meaning of this memory.

> Depitarinja used a sadistic and imperative tone in speaking to the girls. He told them to stand still and not dare to move while he shot and chased them. The girls obeyed him and he shot them with the water pistol. Then he shot the water into his own mouth, calling it milk.

In the screen memory itself the Oedipal situation showed through clearly. In it was the mother who gave the child goat's milk as a substitute for her own milk, the boy who used his knife (a substitute for the penis) on his sister (a substitute for the mother) and was punished for doing so by the father. The pantomime served to emphasize the sexual significance of this memory. His command to the girls to stand still while he chased them (which was obviously impossible) can be understood as a result of his own passive desires in relation to the father. His shooting of the little girls had the same sexual significance as his cutting his sister with the knife. He must be punished for this, just as his father had punished him, and so he turned the pistol against himself. He called the water that he shot into his own mouth "milk." Again, his passivity is clear. In addition, the father was shown as the giving, almost maternal, provider, the counterpart of the phallic mother. The phallic mother complex was the result of the passive attitude of the boy to the mammae which, since they were able to "shoot," must be regarded as a maternal penis. The sadistic attack, followed by masochistic repentance, was a frequent sequence of events with Depitarinja.

> Depitarinja then lay on his back in the sand and mimed an erection and an ejaculation with the water pistol. He referred again to the killing of the bullock by his brother. Then he had the monkey kill the bullock (represented by the goat) by shoving it into the monkey's vagina.

Although the bullock was killed by a man, Depitarinja elaborated the castration trauma as though the threat came from a woman.

> The goat licked the monkey's vagina. Then Depitarinja declared that one of the little girls was a witch. He dug several holes in the sand, calling them caves. He declared that they were inhabited by witches. He placed the water pistol on his penis and squirted the water all about him. Finally, he squirted the water into his own mouth.

The vagina as castrator was represented by the caves, the girl, and the witches.

Hour 23

Nyiki was absent from this meeting. Arnold, a somewhat older boy, took his place.

> Depitarinja saw a picture of an old man in a magazine and identified himself with it. He picked up the monkey and said that it was Nyiki's wife. He then described how Nyiki speared his wife in the anus. He said that the goat was full of feces, which came out of the anus.

The toy's stumpy tail, which had previously been regarded as a penis, he now regarded as excrement.

> He described a picture in a magazine of small winged devils sitting on an umbrella, as Nyiki bending forward and defecating. Throughout this session, Depitarinja was somewhat gloomy and pensive. He looked at the images of the sky, the sun, and finally, himself, in the mirror. The two boys then broke the box of the girl-in-the-box toy. They called the toy by the name of one of the little girls. They mended the cardboard box and then broke it again. They repeated this several times.

Since the doll and the house are both representatives of the female body, this game can be regarded as an abreaction of body destruction and restitution fantasies.

> Depitarinja identified himself and Arnold with pictures of two kangaroos in an Australian illustrated magazine.

I suggested that they play *altjira*, the oracle game. Having collected the necessary objects the day before, I brought them out of the house. There were the curved sticks and the wooden chips which the children sometimes used instead of gum leaves.

> The girls ran to get some real gum leaves while the boys cleared a smooth space in the sand. The chips and the leaves were arranged in a row and hit with the curved sticks. The two chips first represented two women fighting and hitting each other on the head, and then two men who had many wives. The gum leaves represented the wives. The chips and the sticks were placed in a circle, and the rubber doll, who represented the Missionary, performed a marriage ceremony. The sticks and chips were not married in pairs, but rather every object was married to every other object. Depitarinja, represented by the goat, walked beside the Missionary, as a kind of second chief. A quarrel then arose between one chip and his wife (a gum leaf) on the one hand, and the Missionary (the rubber doll) on the other. The Missionary hit both the man and his wife. Then the children stood the Missionary on his head and cried that he had fallen out of an airplane.

HOUR 24

> Depitarinja played with the mirror, staring at his reflection. He said that he saw the image of a man wearing a *totura* (the headdress worn by the non-Christian natives) and a short beard.

From what followed, we may infer that he replaced his own image with that of his father. Before he came to live at the Mission he had been a "free" native and wandered about the bush with his parents.

> He sat in the center of the group and played the role of the father or chief. He distributed the toys in the same way that the chiefs distributed food. He handed the monkey to Nyiki and told him that it was his wife. He gave the snake to one of the little girls and told her that it was her husband. Nyiki held the goat up and said that Depitarinja had put his penis into the rectum of the goat who had then defecated on it. Depitarinja then accused Nyiki of having eaten crow.

The meat of the crow was the only food that an Australian

native would not eat, or at least would not readily admit to having eaten. His reasons were not animistic. The natives regarded crow meat as disgusting. However, in this case the reference was to a real event in which Depitarinja played the principal role. The children had caught a crow in a trap. Depitarinja acted as the chief and distributed the food. Nyiki gathered wood and the little girls prepared the meal, which they all ate.

> Depitarinja began to play with the monkey, singing a little song to himself, "Crow, crow in a hole. I shall catch you." He then decided that the goat was Nyiki's wife or mother-in-law. The goat's nose was placed in the monkey's vagina and the scene was reflected in the mirror. He stated that there was a picture in a magazine of Nyiki's wife. He said that she cohabited with Nyiki while he showed off. Another picture became a demon who was in love with Nyiki, and tried to attract him by showing his rectum. One small rubber doll was given the name of one of the little girls. The goat, her husband, ran away with her on his back. Then the big doll was placed on its back and the little one was shoved into its vagina. The children said that the little doll was licking the vagina of the big doll.

I asked them if one girl could play with another girl that way.

> They said that one girl could lick the vagina of another and mentioned two girls who did just that. The children said that the girls sniffled like stallions when they had their noses in the other's vagina. They said that the two girls loved each other.

This was the last play session at the Mission station. Shortly thereafter I left for the bush.

We know a great deal more about our patients in Europe and America than I can claim to know about Depitarinja, the most prominent figure in the group of children I observed at Hermannsburg. The circumstances of my observation did not really approximate an analytic situation, and the time devoted to observation was too short. Yet, if we look at matters from the point of view of the anthropologist, Depitarinja was in 1932 the best-known primitive child in the world.

His mother died when he was five. His father had left the family group before that time and had gone to live with another woman. Depitarinja was left at the Mission and grew up in the native camp there. He was cared for by one of his classificatory mothers, a woman who cooked for the white teachers, although he actually lived in the boys' house.

I was advised by Mr. Heinrich, the Mission teacher, who knew Depitarinja well, to begin my researches into the psychology of the Australian children with him. He was a bright, intelligent boy, full of life and fun. Mr. Heinrich thought I could get to know him faster than any of the other children, and this proved to be true.

While our play sessions lasted, Depitarinja developed a strong transference relationship to my wife. She was the nourishing mother beloved and feared, while I was the representative of the father. There can be no doubt of his attachment to me. In my field notes of July 5, 1929, I found the following passage: "He always gazes at me in a sort of trance when he comes into my room. In my absence he sits in my chair and looks about him in a pensive way, copying my mannerisms very exactly." While he was able to express most of his feelings toward me in a direct fashion, his feelings toward my wife were displaced onto our kitchen girls, with whom he was continually quarreling.

His aggressive trends were clearly apparent in his play, as they were in the events of everyday life. The phallic snake which symbolized himself or his penis was the center of his play activity. The first scene he acted out with it was the cohabitation of the snake with the woman who nursed it. Depitarinja himself equated this woman with his grandmother.

In his infantile memories, his sister was equated with his mother and played the major role in scenes of antagonism with his father. The family of his games included only son, mother, and sister—the father did not exist. The game in which the white man fell out of the airplane expressed the same wish. At the zenith of acting out his Oedipal desires, Depitarinja was checked by latent castration anxiety. He suddenly discovered that the snake itself, the symbol of the phallus, had no penis. His horror of blood, in describing the killing of the bullock and when he saw my wife kill the chicken, was evidently derived from this anxiety.

In Depitarinja's actual sexual life, the oral perversions expressed by the Aranda words *aruntjima* (to kiss) and *injainama* (to lick or smell) were predominant. Both terms mean cunnilingus, in which the nose, mouth, and tongue are brought into contact with the cunnus or anus. Sexual excitation is derived both from the oral contact and the olfactory stimulus. There was a close connection between this perversion and Depitarinja's castration anxiety. When he spoke of his own experiences with these perversions, he discovered that the snake had no penis. He then referred to the vagina as a wound. These incidents indicate that this cunnilingus is a perversion in the true sense of the word. The anxiety aroused by it is, in part, the fear of the castrating vagina.

Immediately after having made the snake perform cunnilingus with the monkey, he declared that he would tear out the vagina and eat it. Ideas of body destruction played an important part in his stealing. The objects that he stole represented the "good body content" of the mother—not only the father's penis in the mother, but also the maternal phallus itself. The phallic mother concept is a condensation of the primal scene. Depitarinja's dreams of the rabbit with a big mouth who chased him, and of the maternal demon with a big penis, are also representative of this concept. His association following the description of the latter dream shows the connection between the phallic mother and the primal scene (see Hour 2 pp. 81–82).

Stealing, to Depitarinja, meant being a demon, and he identified the demons with clandestine relations with the mother. The demon-phallic mother equation was also prominent in the dreams of the other children. The phallic mother (the pregnant woman) cohabits in the male position. In the play of the children, she cohabited with girls. In his play, Depitarinja obviously identified himself with the girls during the enactment of this scene. These identifications and concepts are typical of the Aranda. Depitarinja developed differently at the Mission station than he would have developed in the bush. At the station there were no initiation ceremonies, no organized male society with a cult of mythical ancestors which could serve as vehicles for sublimation of homosexual trends. In the desert there was a great deal of free sex play between the boys and girls, while at the station the boys lived in the boys' house, where homosexual practices and sadomasochistic perversions were rife. At the station Depitarinja stole continually. However, the situation at the Mission did not seriously harm his good temper, sociability, and the genitality of his libidinal trends.

About Nyiki we know much less than we know about Depitarinja, and we know still less about the other children who participated in the play sessions. Nyiki was less of a "personality" than Depitarinja. He was also closer to the "bush" type. This was to be expected, in that he had come into the station more recently.

The little girls acted with the natural shyness of their sex, with behavior that can best be characterized as "defensive invitation." They were always submitting to being shot. Sometimes, however, they took the male role (this feature of their play will become clearer when we describe the play of the bush children).

Nearly all the dreams which the children told me were about demons. This does not necessarily prove that this was the major content of their dreams. It is quite possible that they thought that the white man was not interested in more ordinary dreams.

The children constantly employed the mechanism of projection in their games. The toys were always cohabiting with each other, and in this way the children obtained their gratification. A favorite amusement of the Aranda youth was to list all the potential wives of their friends. When two of the boys at the Mission were chopping wood, they would mention the name of a potential wife (*noa*) of the other with each stroke. In this game was not only the projection of the desire for coitus, but also a repetition of the primal scene ("He is having intercourse while I am looking on").

Play Sessions in the Bush

The circumstances in the desert were much less favorable for the study of the children than were those at the Mission station. Linguistic difficulties were increased and it was impossible to isolate a small group of children; all who wanted to come were included in the sessions.

Hour 1

I spread out all the toys on the sand in front of the children.

The children had never seen anything like the toys and their first and very natural reaction was to retreat. They recognized the rubber snake as a snake and asked if it would bite. They called the monkey a ghost. The ball was described as a testicle. They decided that the hole in the neck of the rubber doll was where it defecated. They mustered up sufficient courage to inspect the toys more closely. They decided that the snake had two noses and that it defecated through its tail.

Gradually the children got over their first shock. They drew nearer and began to talk to the toys. They treated the toys in a completely animistic manner, and spoke to them as though they were alive. They told the monkey (who was leaning on a miniature tree) that if he did not drop the stick, they would break his arm. When the monkey did not respond to their commands, they began a closer inspection. They mentioned that it had a mouth, hair, and went on to describe minutely every feature of its anatomy. Their chief concern with each toy was to find the place where it defecated. When they found that the goat, which they called a dog, had no anus, they were deeply alarmed. They finally decided that it defecated through its tail. They put forward the same theory with regard to the snake, but without the intervening stage of alarm.

They recognized their images in the mirror. They looked behind the mirror to see if there was anything there. "The monkey [whom they called a dead man] has a pointy anus [a term of opprobrium]," they declared. When they discovered that the monkey had no penis, they decided that it could not be a man. The big doll was called a woman; the tongue of the rubber mask, a water serpent; and the small doll was used as a baby. All the boys pretended that they were giving it milk to suck.

Hour 2

The second meeting took place three days after the first. Pukuti-wara's three sons, Jankitji, Ulurka, and Aldinga, were present. The liveliest of all the children was Wilikutuku, a Nambutji boy about nine years old, who played the same role in the bush that Depitarinja had played at the Mission station.

Also present were Katanka, a left-handed boy, and Tapana, both of whom were Pindupi. There were four girls present—Pantjiti, Witala, her cousin Initnika, and Witwara.

> The children focused their attention on the ape. He was an old man walking with a bent back. They again remarked that he had no penis. They described the monkey's vagina and anus as being stinky like that of a woman.
>
> The monkey had evidently lost some of their esteem. As soon as they had decided that the monkey was a woman, they made both the big rubber doll and the smaller doll cohabit with it.

I then picked up the two dolls and showed the children that neither of the dolls had a penis.

> Wilikutuku finally got out of this difficulty by saying that the paper trumpet was the doll's penis, but that the doll had found it more convenient to drop it.

Wilikutuku had previously declared that the trumpet was a man with a penis. I asked him: "If the trumpet is a man, where is his penis?"

> He answered that *the hollow part of the trumpet was the penis.*

An oblong object with a hollow part can be considered a penis by these people because, in their practice of subincision, a slit is made in the penis.

> Wilikutuku was conscious of the similarity between the trumpet and the subincised penis. He told me that his penis would be like that when he was older.
>
> I asked him if he had ever had intercourse. He denied having had intercourse, but said that he had seen other people performing the act. He then revoked the latter statement.
>
> All the children began to hit the head of the big rubber doll. They said that it was an old man with a bald head.
>
> The small rubber doll was an object of interest because it squeaked through a hole in the back of its neck. They de-

cided that it also defecated through this hole. They pressed it with their fingers to see if any feces would come out. They asked me what it ate, and tried to feed it the dried fruit which I had given them.

They made the snake lick the vagina of the goat, and then cohabit with the monkey. Whenever they pretended that one animal was having intercourse with another, they did it twice; the first time with the vagina and the second time with the anus.

Hour 3

A four-year-old Pitjentara girl, Munyuruna, was present at the third meeting.

> She was interested only in discovering where each of the toys defecated. The snake presented a particularly difficult problem, and she could not settle the question to her own satisfaction. She passed on to an inspection of the monkey. When she discovered that it had a vagina, she made it cohabit with the snake. When she saw that the small rubber doll had only one hole, she decided that it ate flour, the white man's food, and that it used the single orifice as a mouth and as an anus. She declared that the ape was a pregnant woman with a red vagina (a term of opprobrium).

Hour 4

> The children amused themselves at the beginning of this meeting by trying to touch each other's genitals. The game was both hetero- and homosexual and went on promiscuously for some time.

Mrs. Róheim pushed the rubber snake toward Wilikutuku.

> Lelil-tukutu saw this and declared that the snake would bite Wilikutuku's testicles. One boy threw a ball at another. Wilikutuku told him to throw it at the other boy's penis. Tapana, the Pindupi boy, was holding the trumpet and monotonously

repeating "boy's penis." He looked through the hole of the trumpet and decided that that was where it urinated.

I put a record on the gramophone. Whenever the volume increased, Wilikutuku thought that the voice was swearing. He then placed the trumpet (which the children called the penis of a boy or a young man) on his own penis and showed everybody how it looked.

Hour 5

Katana held the ball between his feet and Kankitji tried to take it away from him. Whenever Kankitji pulled at the ball, Katana declared that he was pulling his penis out.

Mrs. Róheim held Aldinga on her lap. His brother declared that Aldinga had a rough anus, and another boy warned my wife that Aldinga might flatulate. One of the boys placed the trumpet on Aldinga's back. He promptly put it on his penis. All the children laughed at this.

I showed the children a picture of some angels in a newspaper.

The children called the angels devils and demons. Aldinga played with the snake and threatened me and Mrs. Róheim with it. Munyuruna took the snake from him and made it cohabit with the goat. Wilikutuku walked about with the trumpet on his penis and demonstrated how people cohabit. He paraded up and down until Ted (an adult) told him to put it into a hole. Wilikutuku instantly told Ted to leave him alone and dropped the trumpet. He again denied that he had had intercourse, but some of the smaller boys told me that he was lying. Wilikutuku and one of the Pitjentara boys walked about, one holding the trumpet to his penis and the other holding the snake to his penis. Still holding the penis symbols in place, the boys fitted them one into the other. The other children protested loudly. There was great excitement. Two girls of about eight grabbed the trumpet and the snake, and using the toys as penises tried to cohabit with my feet.

I was sitting on the sand in the midst of the group of children.

Wilikutuku took the snake away from the girl who was playing with it, and placing it on his penis, thrust it at me, saying that it was going to bite me.

Aldinga sat apart from the rest of the group. He was sulking because one of the girls had taken the trumpet from him. The other children teased him. They shouted that the ball was his testicle and that it had fallen off. Munyuruna said that the rubber snake was a serpent. A somewhat older girl remarked that it was only *like* a serpent. However, the older girl then remarked that the big doll was a demon and threw a stone at it. One of the boys acted out a scene in which the snake came out of Wilikutuku's anus and bit another boy's eyes.

One little boy tried to get Munyuruna to nurse the little doll, but she refused to comply. Several children placed their fingers into the hollow part of the trumpet, calling it a subincised penis. They pulled their fingers out and smelled them. They declared that they stank.

Aldinga placed the trumpet on his penis and walked about the group as proud as a peacock.

Hour 6

Wilikutuku placed the trumpet on his penis and held the rubber ball below it, as a kind of supertesticle. He then put the ball into the trumpet and took it out again, explaining that this was the only way that the semen came out during coitus. He then put the snake into the trumpet and pulled it out, declaring that this was the way that the child came out of the penis.

Muluru, a smaller Nambutji boy, placed the snake on his penis and tried to cohabit with the monkey. He described the monkey as a demon woman and declared that the penis went "right into her." The word he used was *tarpangu*, the word that closes every myth, when the hero "goes right into" the earth.

It is obvious that Muluru felt that this coitus was a kind of supercoitus, because it took place with a demon woman and because of the use of the word *tarpangu.*

> Wilikutuku, in his never-ceasing quest to acquire new genitals, affixed the two rubber balls to his testicles and the trumpet to his penis. He played with my big Alsatian dog. When the dog threw him over, he exclaimed, "Laddie's excrement." The other children imitated the white man. They mimed the way he walked, rode, talked, etc.

Hour 7

> Wilikutuku enacted a coitus scene using the trumpet as a penis. However, he soon dropped the toy and concluded the scene using his own penis. His game developed into a juvenile orgy. He rushed at the other children, both the girls and the boys, threatening to cohabit with them in the normal positions and *per anum.* He also used the monkey and the doll as sexual objects. With both the other children and the toys, he used his own penis rather than one of the symbolic phalli. When he ran toward the other children, holding his penis in his hand, they ran away from him as fast as they could.
>
> I brought out a new India rubber doll. At first, the children treated it as though it were a human being. They described its hat and were particularly interested in the feathers attached to it. Wilikutuku lost no time in using it as a penis symbol.

The children used all the toys as penis symbols. They placed the toys at their genital regions and rushed at one another, threatening to cohabit.

Hour 8

A new child was present at this meeting. He was a thin, weak-looking boy called Muluruna and was brought by Pana's daughter, Pantjita, who was his cousin.

> The children continued the coitus game of the last meeting. Muluruna sat off to one side, holding the trumpet and repeating "boy's penis" to himself. The children told me why

he was so thin and weak. When he and Pantjita had been at Ilpila, he had been bitten by a dog demon. Both children had seen the dog. It had large ears, long teeth, red eyes, and a hairless body. It resembled Pana's dog. When the children had approached it, it had crouched and then it had jumped up and bitten Muluruna. It had not touched Pantjita. Since that time the child had not developed properly.

The dog which resembled Pana's dog was the "demon" aspect of the father's penis. The child may have suffered a shock which mobilized castration anxiety and precipitated an infantile neurosis.

Hours 9 and 10

The children kicked and threw the ball and did not do much talking during both these sessions. Katuna was very skillful with the ball. Muluruna discovered the red apron on the India rubber doll and decided that it was the clitoris. Wilikutuku pushed Aldinga into the fire. When Aldinga hit him, the bigger boy did not retaliate. He satisfied himself by calling Aldinga a "big anus without a hole."

Hours 11, 12, and 13

Aldinga by this time had made himself at home in our tent. The teapot attracted his attention. He touched the spout and then his own penis. He pointed to both and said that they were the same thing. After having received his breakfast, he urinated at Mrs. Róheim. He walked a yard or two away from the tent and defecated, looking triumphantly at my wife as he did so. He then ran up to me and hit me as hard as he could with his fists.

A three-year-old Pitjentara boy put the snake on his penis and rushed at the other children. A Pitjentara girl of about the same age exclaimed that he had an erection.

Tapana ran about with the trumpet shouting, "Penis!" I asked him whose penis it was. Before he could answer, Aldinga claimed it as his own. Before he could grab it, a little girl also laid claim to it. She rushed away in the active coitus position, using the snake in the same way as the boys did.

She described the snake's tail as a clitoris with teeth and the goat and the transparent globe with the fish in it as demons.

There can be no doubt about penis envy in the case of this little girl. She not only used the symbolic penis in the same way as the boys, but also identified the snake-penis with her clitoris. Both she and Aldinga exemplify the attainment of genital primacy at a very early age.

Hour 14

Witala did her best to attract my attention. Her coquettish behavior could not be misunderstood.

My broad-brimmed felt hat had two ventilation holes in the crown. I told the children that they were the eyes of the hat.

The holes were objects of great interest to the children. Wilikutuku thought that the holes were the hat's vaginas, while another boy suggested that the hat had two penises. Aldinga had been bitten by an ant and his penis was very swollen. The children were highly amused by this and shouted that he had an erection. We played with the ball. Whenever I failed to catch it, the children were delighted. Wilikutuku "unintentionally" threw the ball at Mrs. Róheim.

The children saw a camel defecating nearby. One of the girls said that another girl defecated like the camel.

The two figures in the transparent globe had become stuck together.

Witala suggested that they were two demons cohabiting.

Hour 15

The children invented a new game. They used the camels' excrement as a ball. During the game one boy called another a "short penis." The children divided themselves into two groups. The quieter group mimed the act of coitus with the trumpet while the other group played soccer, the rudiments of which I had taught them.

Hour 16

The girls played with the trumpet, using it as a penis. The other children asked me to bring out the "English devil" which "talked" from the box. When the gramophone was brought out, Wilikutuku thought that it was swearing at him. He called it a "big penis." The children thought that the gramophone was saying "Wilikutuku" or "Aldinga." When the machinery "groaned" the children thought that it had been speared.

Hour 17

Witala showed me the trumpet, explained that it was the penis of a big boy, and continued to make overtures to me. A little Pitjentara girl picked up the goat, kissed its anus, and told me with a smile that she was sucking the dog's anus. Tapana held the toy fish to his penis, the Pitjentara girl did the same with the snake while another boy used a doll for the same purpose.

Jankitji, Aldinga's brother, placed the trumpet on his penis and pretended to cohabit with the monkey *a tergo*. He said that the monkey was a demon.

All the children pretended to cohabit. They used the trumpet and tried to cohabit with each other and with me.

Wilikutuku passed from symbolic play to reality. He pinched another boy's buttocks and grabbed his penis. The children cut a hole in the toy goat. They said that it was "the opening of the mouth of its anus." They made another hole in the front of the toy and declared that this was the place where it urinated.

Hour 18

Witala put the trumpet, which the children by then called "the penis of a man" rather than "the penis of a big boy," onto her genitals and slapped me with it. She then picked up the transparent globe and kissed it. She called it "My man." The children usually called the globe "cohabiting," since the two figures in it were stuck together.

I opened the corkscrew part of my clasp knife. All the girls fled in alarm, shouting "Dog's penis!" They listened to the ticking of my watch and described the sound as the one that a stick makes when it is broken.

Hour 19

The children played as usual. Suddenly several of the girls began to shout, "Leave it, leave it," and ran off to someone's rescue. A little fat Pitjentara boy of about three had caught hold of a one-year-old baby girl and thrown her down. He lay on top of her and tried to perform coitus. The big girls pulled him off the little one. The rescuers, however, continued the game among themselves. Witala used the trumpet as a penis and pretended to have intercourse with her cousin. Two of the smaller girls played the same game without any toy to represent the penis.

The actions of the older children were only *games*, while that of the little boy was an actual attempt at coitus.

The children used the camels' excrement as a ball and played soccer with it.

Hour 20

This was the last meeting of the play group in the bush. Very little happened, partly because many of the children's mothers were present. The children became angry when I shared their bread and milk with the mothers. When the women saw the toys for the first time they were frightened, and asked whether the monkey and the snake would bite them.

The leader of the group of children in the bush was Wilikutuku, a good-looking, lively, and friendly youngster. I saw him perturbed only once. I had shown him his own photograph; he shuddered and threw it away, and later refused to speak about the matter. He was a Nambutji boy, and since he was of the appropriate age, he was a passive homosexual, the "boy-wife" of his future father-in-law, with whom he camped. However, this passive homosexual practice had not deprived him of his virile temperament. He missed no opportunity to give the group and myself evidence of his virility.

The only other child in the group who stood out as a character of any importance was Aldinga, the son of Pukuti-wara. He was potbellied, extremely thin, and had a generally poor physique. Far from being oppressed by his poor physical appearance, he was inclined to be domineering. Although his older brothers and the other boys teased him, he usually managed to get his own way.

He was the favorite child of his father, a tall, gaunt man with a rather forbidding appearance, who, as noted earlier, was a great chief, the hero of many war parties, the leader of the great kangaroo ceremonies, and a sorcerer of great fame. The people of the area considered him part demon. The only times he smiled were when he looked down at his small son.

Aldinga preferred the company of his father to that of his age mates. He trotted along beside his father, holding his hand, and was frequently present when Pukuti-wara came to tell me myths that children were strictly forbidden to hear. At these times Pukuti-wara would tell Aldinga not to sit so close to us, but the child rarely obeyed. His father then "growled" at him in the native fashion. When I interceded and told Pukuti-wara that the child could not possibly understand a word we were saying, he was mollified. Aldinga was completely unused to being ordered to do anything.

When his father died, Aldinga's position in the children's group did not alter appreciably, since he had acquired an equally powerful protector in the person of my wife. She pitied the child because he looked so hungry. She gave him all he could eat in the way of canned meat, eggs, jam, bread, and milk, to the great delight of his father and the great envy of the other natives who had children of a similar age. He ate phenomenal quantities. At first we thought this was because he was actually starving, but as his appetite kept increasing, we looked for another reason. As soon as he was given a piece of bread and butter, he would share it with his brothers and his friends. When it was gone he would run back to my wife and ask for more. He was playing the part of the chief, the distributor.

Aldinga, along with some of the other bush natives, followed us back to the Mission, where Aldinga continued to act in the same way. One day, when we had given him some jam, he ran to some old men who were seated with me in the sand, dipped his finger into the jam, and offered each of them a lick.

While Aldinga was constantly playing the chief, he was not above combining the pleasures of this exalted position with those of a very different age grade. At one time, when he had just finished eating a sausage, he saw one of the women nursing her baby. He rushed to the scene of action and demanded the other nipple.

Aldinga tried to get all the pleasure he could out of life, and let nothing stand in the way of his pursuit. To console himself for the fact that he had to stand still while being photographed, he managed to play with his penis and suck his thumb at the same time. His mother, Tankai, did all that she could to please him. One evening when we visited the native camp we saw her playing with Aldinga's penis. He was evidently enjoying the procedure immensely.

He was a remarkably lively and easily manageable child who was never at a loss for an answer. As strange as it seems, he could even behave well by Western standards, and did so whenever Mrs. Róheim allowed him to eat at the table with us. When we returned to the Mission, my wife gave him a little frock. The other children teased him and called him a girl, so he lifted his frock and refuted their lies.

Although Wilikutuku and Aldinga were the two major players in our sessions in the bush, it was not possible to learn as much about them as it was about Depitarinja. This was due for the most part to the large number of children who took part in the play sessions in the bush, and also to the increased linguistic difficulties.

The one fact that emerged with startling clarity about the desert children was that almost their only, and certainly their supreme, game was coitus. This was an important game for the children at the Mission, too. But whereas the Mission children were constantly talking about cohabitation and making the toys perform coitus, the bush children acted out the game, either with their own bodies or with objects held to the proper parts of their own anatomy. The difference between the two groups of children was the result of the influence of the school and the Mission. If we compare these Australian children to European children, there is a striking contrast between the direct libidinal gratification of the first group and the sublimation of the second.

Another aspect of the children's play which interests us is the polymorphous character of the sexuality of the children of the desert. The boys and girls played both male and female roles, with homosexual and heterosexual partners.

In their imitations of coitus they switched from the conventional to the anal position with scarcely any transition. In this respect, their behavior was strikingly similar to that of the monkeys and apes described by Zuckerman.[12] The distinctively human element in this sexual behavior was Wilikutuku's *denial* that he had actually had intercourse. Although he continually pretended to have intercourse, he dropped the symbolic penis when an adult told him to "put it into a hole."

During my play sessions with the children, I introduced toys and other objects that were completely alien to them. I felt that this experiment would show the postnatal accommodation of the human psyche to the environment, and the reactions of the children bore out my hopes. Their first reaction to the unknown was one of fear and anxiety, which they attempted to overcome in an animistic, narcissistic fashion. If they could regard the nonself as being similar to the self, there would be a way to deal with it. They invested the toys with pregenital libido; they had to discover where the toys defecated, where they ate. They finally mastered their anxiety by leaving pregenital solutions behind and turning to more adult, potentially phallic, attempts and expression of instinctual impulses. All the originally dangerous objects became phalli.

The girls used the penis symbols in the same way as did the boys. In their imitations of coitus they played the male role, identifying the penis symbols with their clitoris. At this point, attention should be drawn to the fact that anthropologists have noted a high frequency of male physiques among Australian women.

It is not possible to give any deeper interpretation of the play of the bush children. Such an interpretation may not be as necessary here as it was with the children at the Mission, because less was hidden among these children.

NOTES

1. Carl Strehlow, *Die Aranda- und Loritjastämme* (Frankfurt: Baer, 1908), vol I, pt. II, pp. 80–83.
2. Although Strehlow does not mention it, *aldoparinja* means "belonging to the west wind." The west wind (*aldola*) is believed to be a wind of an especially infernal character.
3. Cf. Géza Róheim, *The Riddle of the Sphinx* (London: Hogarth, 1934).
4. Géza Róheim, "Children's Games and Nursery Rhymes in Duau," *American Anthropologist* 45 (1943): 99–119.
5. Géza Róheim, "Psychoanalysis of Primitive Cultural Types," *International Journal of Psycho-Analysis* 13 (1932): 23–28.
6. Melanie Klein, *The Psycho-Analysis of Children*, 3rd ed. (London: Hogarth, 1943).
7. Because of the way in which our toys are manufactured, the figure lacked genitalia. This lends itself in our minds to the fantasy that it is a female.
8. Unless otherwise stated, the children present were Depitarinja, Nyiki, Tena, Angelica, and no others.
9. Here, as in the remainder of the play sessions, the children kept up a running commentary and description of what they were doing.
10. The children had seen automobiles at the Mission station.
11. Since the introduction of firearms, the Aranda have used the word "shoot" to mean "coitus."
12. S. Zuckerman, *The Social Life of Monkeys and Apes* (London: Kegan Paul, 1932).

5

THE ALKNARINTJA

In Mundane Life

One of the most important people of both mundane and ritual life is the *alknarintja*. None of my informants were clear as to whether the *alknarintja* was an actual woman or a mythical being. The name is used to refer to real women and to characters in the myths and rituals. The *alknarintja* of the myth owned the *tjurunga* in times long past. She owned the bull-roarer as well, which, at a later time, was passed on to the men. She is described as having hair the color of smoke. Her primary characteristic was that she avoided men, but she would respond to the one man who performed a love incantation for her. When he wanted her, she would rush to him from a great distance in a highly excited state.

Old Yirramba told me that a real woman could also be an *alknarintja*. The word itself means "eyes-turn-away." He said that all women become *alknarintja* when they are very small, i.e., they begin with an attitude of avoiding men. Spencer[1] refers to the *alknarintja* as women who will not go with a man unless he has performed love magic for her. She is said to prefer men from distant places. I do not feel that his remarks on the *alknarintja* give enough importance to the significance of the *alknarintja* concept in myth or in everyday life.

Before we can discuss the mythical attributes and their meanings, we must first examine the role of the *alknarintja* in mundane life. I was told that, in the beginning, every woman is an *alknarintja*, and she continues to be one until her resistance has been broken by the man who has "sung" her. The man who wishes to win an *alknarintja* must smear the small bull-roarer with blood taken from his subincision hole. The bull-roarer is swung while the man chants the proper incantation. (The incantations, the *ilpindja*, will be discussed in Chapter 6.) When he has done this, the woman sees his *altjera* (ancestral spirit or double) penetrating into her body and calling her by name. She, who until this time had nothing to do with men, now falls in love. She continues to resist the advances of all other men and will give herself only to the man who has "sung" her.

The Pindupi and the Yumu call an *alknarintja* an *aninpa*. She lives in the girls' camp. The men who live nearby all advise their friends to take her, but the girl takes great pains to hide herself from the men. When she leaves the camp she walks on the grass in order to leave no track. At night she does not stay in the camp with the others, but erects a windbreak in the bush and stays there. In the morning, when she goes to gather seeds with the other women, if a man should approach her she runs away and hides in the bush. Finally, when the man who has made her into an *alknarintja* appears, she succumbs to him, and he takes her by the arm and marries her.

From this description we see that the *alknarintja* may be a woman who has resisted the advances of men since she was a baby, or she may have become an *alknarintja* after having been "sung" by one man. Although every woman may be an *alknarintja* to begin with, it must be noted that there are no spinsters among the Central Australians, and therefore no woman lives and dies an *alknarintja*. It should also be noted that all the men with whom I spoke had had intercourse before the age of puberty. Nonetheless, the *alknarintja* attitude of the woman is not merely a conventional facade to be cast aside as soon as the opportunity presents itself. The custom of *mbanja*, which is directly related to the *alknarintja* concept, often requires real force on the part of the man and real struggles on the part of the woman and her friends.

Strehlow[2] has recorded the myth of the Alknarintja of Iloata (Mt. Conway), which illustrates the relation of the *alknarintja* concept to the custom of *mbanja*. A group of *alknarintja* women met a man of the Little Hawk totem and asked if he were related to them. Replying that he was their husband, he put his spear thrower under his arm in the typical *mbanja* attitude and tried to drag off one of the women. She bit his hand until he released her; then she and her friends hit him with their yam sticks.

The *alknarintja* is connected in some fashion with the small bull-roarer. Motna, a southern Aranda, told me that the small bull-roarer was originally owned by an *alknarintja* called Djirindjira (Grass-woman), who later gave it to the men. He said that when a man wishes to charm an *alknarintja* he throws the bull-roarer in her direction and then pulls it back toward himself. The woman then dreams of the man and embraces him. The same thing occurs when a woman conceives.

Having seen the role that the *alknarintja* plays in mundane life, we can now focus our attention on the *alknarintja* of the mythical period. The following collection of myths and songs has to do with the *alknarintja*. They are used at various totemic rites. In those cases where I saw or obtained a description of the corresponding ritual, that information is included.

The Myths and Songs of the Alknarintja

Song and Myth 1

The following song of the Honey Ant totem was given to me by Old Yirramba. The song, and the myth which corresponds to it, are used at a totemic rite at which a ceremonial pole is erected.

THE ALKNARINTJA OF AROULBMOULBMA[3]

The alknarintja woman is tired.
She refuses to move.
She sits very still.
Near her is an ants' nest and a ti tree bush with its branches stuck together.
All the alknarintja women are tired.
They sit down and refuse to move.
Near them are erijila flowers and a tnyelinga bush.
They are sitting near a big camp.
They sit on the flat sand.
About them grow many tnatata flowers.
The pollen of the flowers fills the air.
They bend down the branches of the mulga tree and pick its flowers.
They pick the tnata flowers.
The branches of the mulga tree are laden with flowers and bending with the weight.
Beyond where they sit is a narrow gap in the hills.
Beyond the gap are many camps.
The ants crawl on the ground.
They look like decoration marks.
They look like decoration marks.

One of the concepts found frequently in the mythology of the *alknarintja* is introduced in this song. The *alknarintja* women are motionless; they refuse to move. Not only are the women immobile, but the objects in their environment are also motionless; for example, the branches of the bush are stuck together and so the bush cannot move. The burgeoning fertility of the bushes stands in sharp contrast to the rejecting attitude of the *alknarintja*. The following myth is related to this song.

> Many *alknarintja* women of the Honey Ant totem lived at Aroulbmoulbma. They collected undjiamba flowers, brought them back to their camp, and soaked them in a bark vessel. They also found red flowers and mulga flowers. They went out looking for the flowers of the iron-wood tree and for *tjuarka*. All the flowers were soaked in water, which the women drank. They then went to a place called Ipmana-pundja (Black-piss-ant), where they found black piss ants. Again they soaked mulga flowers, *tjuarka*, iron-wood flowers, ti tree seeds, and the branches of the witchetty bush in water. Then they returned to Apma-kapita (Snake-head), where they gathered the flowers of the ti tree bush and took them to the ancestral cave. They went out again and found many different kinds of flowers. They went back to the ancestral cave, where they became *tjurunga*. They were Pananga and Pangata women. They owned ceremonial spears and decorated themselves all day. Their leader was Tnatjilpaka (Stand-up). She stood up at Indotapa (Boomerang) and became tired. A gum tree stands there now.

Song and Myth 2

The following myth and song were related to me by Aldinga of the Merino group. The Merinos are a group of Lurittya of the type that would be called "half-Aranda" by the natives. The song is used at a multiplication ceremony which only the initiated are permitted to attend. The performers make a hard surface of the sand by permitting blood from wounds made in their subincision holes to drip onto it.

The men make marks in the sand and then toss the sand into the air. The old men sit motionless, holding their shields and wearing their large *tjurunga* ("body" *tjurunga*) on their heads. They eat euro meat before singing the song. The ritual ensures the reproduction of the euro.

The two men are wandering in search of euro.
They see a woman far off on the road.
They cook the fat of the euro in its skin.
They strike the erultja bush with their ceremonial spears.
Near them is a red salt bush.
It is bad, for salt bushes should be blue.
They make spears from the branches of this bush.
The spears are crooked.
They do not want to hunt euro now.
They want to hunt women.
They see the nest of the arkana bird among the rocks.
As they run along, they decorate themselves and put on their arm strings.
They dance. Their feet thud on the ground.
Their feet thud on the rock where the bad spear bush grew.
They think to themselves, "We will make a spear."
"We will make a spear that the women may know we want them."
They see the women standing in a row.
Again they see the arkana bird.
It reminds them of the old blind woman back at the camp.
The bird's cave is black.
It has been burnt black by the bird's fire.
The men sit down in a dry bush.
The bird flies away.
They hear the women approaching.
The men dance on the rock.
Their feet thud.
They dance on the ground.
Their feet thud.

Before we attempt an interpretation of this song, it will be necessary to relate the myth associated with it. The myth differs from the song most strikingly in that there is no mention made of the women (presumably *alknarintja* women) whom the men see in the song.

An old woman and two men lived at Alknarintja-arakutya-indara (Alknarintja-woman's-vagina). She was their father's sister. The older of the two men gave her kangaroo meat, but the younger did not. She would ask him, "Have you killed any euro?" and he would answer, "No." She was blind and so she could not see what the younger man was doing.

He left the camp one day and went to the southwest. There he killed two kangaroos. He brought them back and cooked them in the camp. Then he fell asleep beside the fire. The elder brother went to get some eagle down with which to decorate himself. The old woman smelled the meat. She took it out of the fire and ate it. When the elder brother returned to the camp, his body was covered with eagle down. When the younger brother saw him, he shouted, "Wah! Wah!" and danced around him.

Each of the brothers made a ceremonial spear. They placed them so that they formed a cross. There is a big stone there today to mark the spot where the two spears stood. The brothers then went to the north where they killed an euro. They found a big white rock on which there were many euros. The younger brother speared the euros and cut them up on the spot. He took only the fat back to camp to cook. The older brother went to the south where he found two euros which he brought back to the camp. Then the younger brother went back to the place where he had hidden the euro meat. He dug it up and brought it back to the camp where both men cooked their meat.

The older brother gave some of his meat to the old woman, but the younger one gave her nothing. Neither of them gave her any fat, but she smelled it cooking. "Why don't you give me any fat?" she asked. Even the elder brother had never given her any fat. Then she made many *tjurunga*. She decorated herself and put the *tjurunga* on her head. The two men went back to the cave. There they found a great deal of eagle down. They all became *tjurunga* together in the cave.

There is a variant of this myth.

> A blind *tneera* (this word means "beautiful woman," but it is also another name for the *alknarintja*) of the Euro totem lived with her two nephews. The two men were good hunters, and always had success when they went to spear the euro. The two brothers would never give their aunt any fat. One day, however, the younger one gave her some caul fat. She rubbed this on her eyes and was instantly cured of her blindness. Thereupon she spat on the euro bones which lay about the camp. The bones were immediately transformed into live euros which sprang up and began to jump about. Previously, all the euros had been blind and therefore easy to spear, but now the euros could see. When the nephews heard the thuds of the euros' feet upon the ground, they knew what the old woman had done. They were very angry that she had spoiled their hunting and tried to kill her, but instead, she seduced them. After they had had intercourse with her, she taught them how to kill the euros with their spears and with black magic.

These myths are used in ceremonies which ensure the multiplication of the euro. "To gain eyesight" is a euphemism for "being born" in several Central Australian languages. A statement that the young kangaroos are jumping about and have gained their eyesight really means that they have jumped out of their mothers' wombs (or pouches) and have been born. It should also be noted that the Central Australian youth frequently refer to their mothers as *banga* (the blind one). Since multiplying one's own totemic species is symbolic incest, we need only rearrange the sequence of events in the myth just related to make it perfectly comprehensible. The nephews first have intercourse with the old woman, i.e., they have incestuous coitus. She is thus made pregnant and gives birth to the young of the species. Since the Central Australians are among the few peoples of the world who regularly eat their own totemic animal, the old woman is able to teach the young men how to kill the euros. Nonetheless, there is a difference between killing the totemic animal and killing any other animal, and so they must rely upon black magic.

The first myth and the song differ from the second myth in several respects. In the song, the old woman (as a sexual object) has been transformed into the *alknarintja* women, whom the men are trying to win through the use of the *ilpindja*. No mention is made in the song of their having sexual intercourse with the *alknarintja*, and in the myth there is neither the transformation of the old woman into an *alknarintja* nor mention of coitus with the aunt. All overt sexuality has been repressed. It is obvious that the *alknarintja* is a distorted form of the original incestuous love object. Lest there be a breakthrough of the repressed, mention of both coitus and the multiplication of the species must be omitted. The *alknarintja* is the representative of the incestuous love object.

Song and Myth 3

I learned the myth and song which follow from Mulda, Wapiti's classificatory brother. They are used in connection with a "making" ceremony, although it is not quite clear as to what they are intended to "make." In my notes I found a statement to the effect that the wombs of the women open during this ceremony, from which one can conclude that the song and myth are intended to "make" love. The meaning of this song and myth would therefore appear to be approximately the same as that of the other song and myth I learned from Mulda. However, at the same place in my notes I found another statement that the *alknarintja* menstruate a great deal, so this ceremony may be intended to "make" menstruation.

The alknarintja are menstruating.
The blood flows out.

They tie their arm strings tightly to stop the flow.
As they walk, they pass a ti tree bush laden with flowers.
They see a tjora bush bent under the weight of its seeds.
The branches are hanging down under the burden of their seeds.
They sit down near a lupulupa bush.
They cover their vaginas with their heels.
Far off they see the track of a man.
It is a clear track and far away.
Within their bodies, they hear the growling of the bull-roarer.
The bull-roarer talks.
The bull-roarer thunders.
"It has entered into me and become a child in my womb.
"My womb has opened at the sound of the bull-roarer.
It has entered into me and become a child in my womb.
"I am in love."
The bull-roarer talks loud.
The man of the Iwupa Worm totem stands with his tjurunga tied to his back.
He is excited.
He bites his beard.
He whirls the bull-roarer.
The string breaks!
The bull-roarer whizzes through the air.
The women stand, holding their yam sticks.
The bull-roarer talks!
The bull-roarer talks!
The ground is cold.
A cool wind blows.
The man is excited.
His Adam's apple tightens.
The women rise up from the grass.
Again the bull-roarer talks.
The man's Adam's apple tingles.
He has performed an ilpindja.

It is morning again.
The heat rises.
The sun rises.
The alknarintja women walk softly.

They have decorated themselves.
The pointed antara bush grows by the road.
Its flowers smell foully of urine.
The women sit quietly, playing the oracle game.
They sit still on the ground and hit the leaves.
They sit in one place and hit the leaves.
They tear up the branches of the bushes.
They make a soft ground for the camp.
They blow out the fire.
One woman puts on her head band and walks quietly into the scrub.
"Where are you going?"
"I have taken everything I own.
"I want to go back to the men."
The other alknarintja beg her not to go.
Her eyes are blurred with tears, but the feeling in her womb draws her on.
The others remain, seated about the dead fire.
Their wombs itch.
Their menstrual blood has ceased to flow and their wombs itch.
They sit in a circle.
In one place, without moving, they sit together.

This song contains, as did the previous ones, a description of the immobility of the *alknarintja* and the contrast between it and the overripe fertility of their environment. A new element, common to many of the *alknarintja* songs, is introduced. The *alknarintja* are menstruating and therefore unable to have sexual intercourse. At the same time, they are extremely excited. As we continue with our presentation and interpretation of the *alknarintja* data, it will be seen that almost every statement made about them is followed almost immediately by its exact opposite. The reason for this will be explained later. The following myth is related to this song.

Two *alknarintja* women lived together at Aningitt-ja-ampingi (Walk-walk-softly). They had a baby.[5] Everyday they left the camp to look for food and witchetty grubs. When they returned, they erected a big ceremonial pole and performed ceremonies. One day, they went to the southwest and found many witchetty grubs in a big tree. They returned to the camp and decorated themselves for the ceremonies. Then they went to another tree and found more grubs. When they returned to the camp, they worked on the ceremonial pole. Then they went to the northwest to find more witchetty. When they returned to camp, they worked once more on the ceremonial pole. Then they went to hunt for opposums and when they had found three they went back to the camp where they erected several very high ceremonial poles. Then they went to the east where they found more witchetty. When they returned they made a short ceremonial pole. Then they went north for more witchetty and when they returned they performed an *illapangura*. They left again, this time going south, and when they returned they erected a very high ceremonial pole.

Then they went to the northwest and there they saw men of the Mosquito and Iwupa Worm totems. These men were in the classificatory relations of son to their child. One of the men had killed three euros. He gave two to the women and kept one for himself. The two women made a camp and the man camped nearby alone. There was a pool between the two camps. The women continued to go out in search of witchetty grubs and perform ceremonies while the man hunted for euro. Then the women and the man felt very tired. The women went right into the earth and a fruit tree marks the spot where they went in. The man also went right into the earth and a ti tree marks that spot.

Song and Myth 4

The myths and the song that follow belong to the tradition of the Honey Ant totem. The two myths, which are used in conjunction with the one song, are included because they add significantly to our understanding of the latent meaning of the *alknarintja* concept. The song is sung at the totemic ceremony that forms the last part of the initiation rites. Ceremonial poles are employed at this ceremony, and the men wear bark vessels on their heads in place of the more customary *tjurunga*. In accompaniment to the song, the old men and the youths dance, circling around the ceremonial poles. However, the myths are also connected with the fertility rite of one class of the totem. Instead of the shields used in most fertility rites, bark vessels, called *tmara*, are used. This rite is performed at Ljapa. Only the old men are allowed to witness it. They sit on the ground and sing, but they do not dance in circles around the ceremonial poles. The song of the old men will be given later.

Yirramba and the young men walk in a row.
They walk in a straight line.
They hear the alputakaputa bird, the familiar of Yirramba, talking.
Yirramba and the young men make ant holes in the earth.
On either side of the road they make holes.
The alputakaputa bird talks again,
And Yirramba and the young men make more holes.
Like ants, they creep up to the mulga flowers.
The leaves of the mulga are dry.
The young men, the ants, crawl into the mulga bush.
They have the vessel, used for collecting ants.
The alknarintja women go into the bush with them.
There they erect a ceremonial pole.
The ceremonial pole keeps the alknarintja women from running away.
Their Adam's apples tingle.
Their penises become erect.
They have kept the alknarintja from running away.
The ants scratch the sand.
They throw the sand about.
They make a big noise.
This is the home of the ants.
This is the bark-covered tree in which they live.
Again they make a big noise.
They return to the place from whence they came.

They sit down.
They look like iwupa worms with long eye lashes.
The little holes of the ants' nest are empty.
In the big hole is the tjurunga.
The young men, the ants, carry the tjurunga about with them.
They carry it in their armpits as they hunt for mulga flowers.
Ilpmana, the alknarintja, the mother of the honey ants, goes everywhere with them.
They come to a camp with a fire.
By the great light of this fire they sit down.
They sit down all together in the men's camp.
Their tjurunga are covered with the bark of the root of the mulga.
They carry their tjurunga with them as they wander.
They carry their tjurunga with them as they go right into the earth.
There they find the dry leaves of the mulga.
Ilpmana, the alknarintja, goes right into the earth with them.
The letipa bird, the father and the chief of the honey ants,
He goes to the north with his sons and the alknarintja.
They go away to the north.
They walk in a line.
They walk in a straight row.
Along the Ljapa road they make their camp.
The father and his sons are together.
The alknarintja woman camps apart from them.
They have come a long way along the road.
Now they sit down.
Along the Ljapa road, as they wandered toward Ljapa, they saw the smoke of many campfires.
There were campfires all around them, but far off.
And so they grew sad.
Their tracks in the sand have been hardened by the rain.
They make new holes in the earth.
They throw the sand all about.
They make the little holes of the ants' nest.
They make the holes in a row.

In this song the realistic and the mythical elements are welded together. Ants are particularly suitable objects for symbolizing the life of the ancestors. The ants dig holes in the earth and live underground, and the ancestors "go right into the earth" and find their eternal resting place under the ground. All the uterine symbols—the holes in the ground, the caves in which the *tjurunga* are stored, the bark vessels in which ants are collected—are used in the song and ritual.

Perhaps the most noteworthy feature of the song is the fact that the *alknarintja* woman is the mother of the totem. She partially fulfills the wishes of the sons: she abstains from intercourse with the father. The following myths are related to this song.

> Yirramba was the greatest chief of a place called Kourpula (Black-head). The young men who stayed there with him made a big mound of sand and performed totemic ceremonies on it. Then they made another mound of sand, a long one. They left the camp and a short distance away came to a big swamp called Etuna (Sand). They sat down and made a camp. There they performed ceremonies. Then they went and camped at Ipmanapuntja (Ipmana-ceremony) near a big water hole. Then they went to a place in the west called Altuna and finally to a place called Wotta (Swamp). There they decorated themselves and again performed ceremonies. They left Wotta and went to Erkirinja (Sores).
>
> At Erkirinja, Yirramba scratched at the sand all day until he had scratched a round vessel in it. He continued to scratch at the sand as he walked to a place called Arumbia (Clay-pond). He carried the vessel of sand on his head like a *tjurunga* until he came to a second place called Arumbia. Then he and the young men went to a place called Engua (Cave). He went away again, continually scratching at the sand and fashioning it into vessels. He went to another place called Engua where there was another chief of the Honey Ant totem and his young men. The two groups united and performed ceremonies. Yirramba was very white. The other chief and his young men came to see Yirramba and the vessel of sand which he carried on his head. Then the second group went away. Yirramba kept scraping at the sand until he went

halfway down into it and came to a place called Ngurra (Place). Then all the young men went into the sand. There was a big camp inside. Then Yirramba went right down into the earth and became a *tjurunga.*

Yirramba made a big ceremony in a big camp in a place called Arumbia. The initiates ran around him in the ceremonial dance. Another group came and joined them and they all went to a second place called Arumbia. There was a big ceremony going on there in which ceremonial poles and bark vessels were used. Yirramba and his young men made a cave in which to perform their ceremony. There they stayed and did not wander away. They all became *tjurunga* in a stone and went right into the earth. They descended on their ceremonial spears. The whole group, with their pubic tassels and their bull-roarers, hung onto the ceremonial pole. Wotta was the chief of the place to which they came. He lived under the ground. The sand vessels went right into the earth and they too became *tjurunga.* They are all still there.

These myths are used at a fertility rite as well as during totemic ceremonies. In the former instance, the following song is used with them:

This is the big place where the honey ants are made.
They turn around in the sand.
It rains when the honey ants are made.
And the toadstools grow.

Song and Myth 5

I learned the following song and myth from Wapiti, and also saw him and his friends perform the ceremony associated with it. Before they began the ceremony they wound string around a yam stick. They let blood pour from small cuts in their subincision holes onto the string, and then affixed charcoal and eagle hawk down to it. They dug a small hole in the ground and let their blood pour into that. Eagle hawk down was placed in the little pool of blood, and then a ceremonial pole was stuck in the hole. The pole (or spear) was decorated with the feathers of the night hawk.

Mulda and Ilntjirilka played the roles of the two *alknarintja* women. They wore head rings with *alpita* tassels hanging down about their ears. Their faces were decorated with white paint and eagle hawk feathers. On their chests were painted circles which represented the breasts of the *alknarintja* women. They circled around the ceremonial pole on their knees, appearing almost to glide. As they crawled, they brushed away the sand with boughs held in their hands. I was told that the ritual represented the *alknarintja* women cleaning themselves after menstruating. The gliding movements were frequently interrupted as the two men got to their feet and ran around the ceremonial pole in the usual ritual manner.[6]

Looking out over the vast reaches of the country,
The hawk sings.
The alknarintja women cut their breasts.
On their breasts they make scars.
They slap their thighs.
They strike the ground and the seed bush that grows there.
They walk about, stepping on the hard ground.
They blow their noses.
As they pass the seed bush, they stamp their feet on the ground.
A rat jumps up and frightens them.
They see the white leg of the kangaroo.
They see the kangaroo's leg, the white leg of the kangaroo.
They walk over many hills, past many bushes.
They carry their yam sticks.
Now they are tired.
They sit on the ground.
"My sister has a red string," says one of the women.
"My sister has a red string."

They stand on a sand bank.
The sun is very hot.
A man, sitting at the top of a big tree, watches them.
At Uralpminja they make a camp.
They are menstruating.
Their flanks are wet with blood.
They talk to each other.
They make a bull-roarer.
They split the wood to make this bull-roarer.
The bull-roarer talks.
As the bull-roarer talks, another answers it.
Alas, the bull-roarers have split!
Their sound is louder.
The bull-roarers thunder.
The bull-roarers talk.
One talks from here, the other answers from there.
Alas they have both split!
The bull-roarer of the woman of the Pigeon totem talks
incessantly.
The alknarintja women clean their camp.
They make a hard place on the sand.
They raise dust all about them.
The baby cries and cries.
He walks on all fours.
He walks holding his little stick.
The women don their arm strings.
They throw sand about them.
They sit down.
They are menstruating.
The blood is perpetually flowing.
A tall ilpila tree grows where they sat.
The red string is bad.
The strings are tied cross-wise.
At Uralpminja the alknarintja women make a camp.
There are leaves on the ground.
There are feathers on the ceremonial pole.
Take them down.
They cut a straight bush.
Take them down.
They go to the north.
Take them down.

This song, like the previous ones, contains many of the typical *alknarintja* motifs. Again there is the ambiguous statement, "The bull-roarer splits." In this song it probably refers to castration anxiety, rather than to the manufacture of the bull-roarer. From the mention of the crying baby it would appear that the women have conceived and given birth. The sound of the bull-roarer entering into their wombs caused them to conceive. In the myth related to this song, the women are again the owners of the bull-roarers. A man is introduced, but the women will have nothing to do with him.

> There was a big camp of *alknarintja* women at Uralpminja (Ashes). In the morning they went out of their camp to find seeds, and when they returned they whirled their bull-roarers. One day, when they had returned to the camp, they divided themselves into two groups. One group went to the west and the other went to the east. The leader of the first group was a woman called Mutta (Little-rat). While she was gathering yams, an euro saw her. The euro was frightened and ran away. Then she and her women went back to Uralpminja. From there they went to Aningi-tjampi (Short-ceremonial-spear), where they found other *alknarintja* women who whirled their bull-roarers day and night.
>
> The group that had gone to the east went right down into the earth as they kept whirling their bull-roarers. The strings attached to the bull-roarers were broken as they did this and the bull-roarers flew into the air and landed at Ltalaltuma. The bull-roarer of a woman of the Pigeon totem flew to Aningi-tjampi. There, the women who found it made a ceremonial spear of their yam sticks and whirled the bull-roarer all day long.
>
> A man of the Opossum totem came to their camp and tried to take one of the women, but he could not get her to go with him. He then went to Atunguma (Guts) and found several women there, whom he took to Ipmilkna. The *alknarintja* women at Aningi-tjampi and at Uralpminja became *tjurunga*.

Song and Myth 6

The next song and myth were used at the next to the last rite of the initiation ceremony. The uncle of the initiates and all the women of the group dress themselves as *alknarintja*. They wear head bands and arm strings. While the women who have dressed as *alknarintja* act as spectators, the initiates, one by one, climb to the top of a ceremonial pole, and each takes one of the *tjurunga* which is hanging there. They then turn around and climb down very quickly. This maneuver is repeated several times. When the uncle has finished singing the *alknarintja* song, he takes the head bands and arm strings from the women and gives them to the boys, who wear them from that time on.

The alknarintja sit with their smoky heads turned away from the man.
As they stand up, they hold their yam sticks to show that they will not go with him.
As they sit, they wear their head rings.
As they stand, they hold their yam sticks.
They decorate themselves.
They wear their nose bones.
They whirl their yam sticks, their bull-roarers.
Alas, they have split!
The bull-roarer, the loud singer, splits.
The yam stick, the long bull-roarer, sings.
The bull-roarer sings.
They stand by a tall fig tree.
The young tree has grown.
They say, "I won't go with you."
"I will remain an alknarintja."
They whirl their bull-roarers.
They stay where they are.
They sit very still.
The man wants them to say, "I will go with you."
But they remain where they are.
They hear the bell bird, the messenger of love.

The bell bird talks.
The women, the fat lovely women, remain seated.
They have gathered seeds.
They rub them in their vaginas.
They wear their head rings.
The bull-roarer talks.
The bull-roarer talks.
But they sit still and avert their gaze.

Of all the songs that we have seen, this is probably the clearest and easiest to understand. In any myth, an *alknarintja* may be recognized by the fact that she is constantly decorating herself with red ochre, that she wears a head ring with *alpita* and arm strings, and that she is usually associated with water. The bell bird, which tries to take the women to the men, appears when an *ilpindja* has been performed. The following myth is associated with this song.

Many women of the Alknarintja totem who were themselves *alknarintja* and would have nothing to do with men lived at Ilpila. They belonged to the Purula and Kamara classes. They collected many seeds, but since they had no water with which to prepare them, they were forced to use urine. The next day, they again found many seeds and also some guanos and rats. Again they prepared the seeds by urinating on them. This went on for several days.

A man called Patu-walanpa (Hatchetman), a Purula of the Alknarintja totem, decorated himself with a nose bone, a head band, and a white pubic tassel made of the skin of an euro. He saw one of the *alknarintja*. She was a Kamara and a very pretty one, a fine fat woman. His penis immediately became erect. He caught hold of her, knocked her down, and had intercourse with her all day long. Finally, he pulled his penis out of her vagina, sat down in the shade, and went to sleep.

In the morning, the man went to hunt for euro and the women went to gather seeds. In the evening, they all came back to the camp. The man wanted the same woman and he tried to pull her along with him, but she resisted and would not go with him to his camp. The *alknarintja* women had found a soakage. They drank some water and rubbed the seeds that they had found with the water and ate them.

The next day, the *alknarintja* women found some seeds and guanos while the man killed an euro. All that day, he had been watching the woman he wanted. Finally, he went over to where the women were standing, took the one whom he wanted by the arm and dragged her off. He pulled her right into her camp and cohabited with her then and there in the presence of all the others. He had intercourse with her all night long until the daylight came.

In the morning, the women went to the east for seeds while the man hunted euros. In the evening, the women returned to the camp with many seeds. Then they climbed a big rock where there were many caves. The man joined them and they all whirled their bull-roarers the whole night long. They erected a big ceremonial pole and fastened their bull-roarers to it. Then they all went into the cave and became *tjurunga*.

This myth and song are particularly significant in that they form a part of a ceremony in which the women participate and in which they are permitted to see *tjurunga*. The women, who represent *alknarintja*, are the owners of the objects which the young men are given as symbols of their manhood. The fact that the *alknarintja* was the original owner of the bull-roarer and was considered to possess a penis is certainly the determinant here. Whether the participation of women in the rite is a survival of an earlier form or the result of the breakthrough of the original idea cannot be ascertained.

Song and Myth 7

Many alknarintja women walk about leaning on their yam sticks.[7]

All of them carry their tjurunga with them as they walk about leaning on their yam sticks.

Finally, they sit down.

They had passed around a dense scrub carrying their tjurunga on their heads.
The yalka grew so thick that they were like a fig tree.
Suddenly the women stand up and divide into two groups.
They shout to one another.
It is raining and very cold.
The women cry because of the cold.
The rain falls.
One old alknarintja woman begins to sink into a boggy place.
She scratches at the earth trying to get out.
She sees a duck as she struggles.
The duck has a big belly.
It sits down near the women.
The smoky haired alknarintja dig for yalka.

This song repeats the motif of the bushes being so thick and so pressed against each other that they cannot move separately. Again the association with water, this time rain, is present. The duck with the large belly is undoubtedly symbolic of pregnancy, but its exact meaning in this song is difficult to ascertain. In the myth associated with the song, the meaning of certain elements of the song becomes clearer.

In the Tjoritja Aranda country there was a place called Indankangua (Seed-place). Many *alknarintja* women of the Honey Ant totem were staying there. They were of the Paltara and Pananga classes and they had originated there. They gathered indanga seeds and yams which they ground up and ate as soon as they found them. There was a water hole nearby in which a kunia snake lived. In order to find water they set out many times for other places, but they could never get away because they were surrounded by lakes and there were many swamps in which they got stuck. They would start out, be forced to return, and then start out again in search of another gully. Then there was a big rain which caused a flood. The *alknarintja* women went back to the place where they had originated and became *tjurunga* there. They were all covered by the flood.

It is apparent from the mention of the snake that the rain is symbolic of semen. This would explain the pregnant duck.

Song and Myth 8

The following myth and song were told to me by Mulda. They, and another, shorter song related to them, are used at the Mbatjiaalkatjuma ceremony. They are performed in order to arouse the participants to the extent that they will have intercourse with anyone present at the ceremony, including those forbidden to them by the kinship regulations. The men taking part in the ceremony rub little stones against a rock and "sing" them:

> Lumps in the flesh of he who cohabits wrong.
> He who cohabits wrong.

Porcupine grass is stuck into a rock by a man who then kneels and goes through the motions of coitus with the rock. The rock is called the Iwinjiwinji (Mosquito) rock. This rite is also performed as part of the Nankura ceremony.

> The ground is cool.
> The ground is cool where the alknarintja sit.
> A feeling in their wombs had drawn them on toward their eternal camp.
> Now they know that they are at the right spot.
> They know because of the feeling in their wombs.
> They walked all in a row and saw a "beautiful man."
> The women tremble with desire for him.
> The ground is cool.
> A breeze blows.
> But the alknarintja, the committors of incest, tremble with desire for the man.

They are hot with desire even while the cool wind blows over them.
Their wombs tremble, these committors of incest!
They cannot sleep.
They lie awake and think of the man.
O the man of the Mosquito totem!
O the man on the road!
Now the women and the man go right down into the earth.
They become tjurunga.

The "feeling in their wombs" which drew them on was caused by the *ilpindja* which the "beautiful man" had performed for them. The magic drew them to the spot where the man waited. Instead of the coitus which usually follows the performance of an *ilpindja*, the man and the women "go right down into the earth" and join the ancestors.

A man of the Mosquito totem, from a place called Pangatuma (Hit-dry-grass), went to a place called Akantjirkni (Ceremonial-pole) where he erected a ceremonial pole. Then he went to Yutarinda (Porcupine-grass-sleep). There he made a windbreak of porcupine grass and went to sleep. From this place he went to a place called Inimba (Semen). There he saw two *alknarintja* women. He was urinating as he watched them and his semen came out.

One of the women stood in the relationship of wife to him and the other in the relationship of mother-in-law. He followed the urine and their track until he returned to Pangatuma. There he found an empty camp. He picked up their track again and followed it until he came to a water hole called Tulpurta (Salty-ground). From there he tracked the two women to Ngunmurkna (Gum) and then to Indutakungura (Normal-healthy-place). There he found them.

He was a normal healthy man. He had intercourse with both of them, first with the woman who stood in the relationship of wife and then with the woman who stood in the relationship of mother-in-law. Then they all went to Aningitjampi (Walk-step) and from there to Itjiti-tjita (Bush) and finally to Iwinjiwinji (Mosquito). Then they all became *tjurunga*.

SONG AND MYTH 9

This song and myth are part of the ceremonial rites of the Allaparindja totem. They were related to me by Old Yirramba, whose father's mother belonged to the Allaparindja totem. The women referred to are often called *labarindja* women. I was told by my informants that a *labarindja* is a real woman who is "wild," i.e., who refuses to have anything to do with men. They are described as having blue skins and smoky colored hair, which are the characteristic features of the *alknarintja*. However, the *labarindja* are also considered demons. Old Yirramba told me that if a man had intercourse with a *labarindja* woman he might die, because the *labarindja* have evil magic in their vaginas. The ceremony in which the song and myth are used consists of the usual dancing and singing. The old man who represents the *labarindja* in the ceremony dresses like the woman in the myth.

The Allaparindja women sit on the gravel, on the hot ashes.
They sit very still on the hot ashes.
Their vaginas look black.
Their vaginas look black.
They wear the feathers of the black cockatoo.
They are shy.
They say, "We two are shy."
They put their seeds into their vaginas.
They put their seeds into their vaginas.
They are fat seeds, good seeds.
The women sit down.
Their vaginas are pointed to the labia.[8]
They play with two little sticks.
They strike them together.
"We hit them," they say.

The *labarindja* have most of the attributes of the *alknarintja*. My informants considered the most important characteristic of the *labarindja* the fact that a man was sure to die after having coitus with them. At the beginning of this chapter we mentioned that there is a tendency to regard every young girl as an *alknarintja*. A conclusion may be drawn that the Central Australians regard coitus as an extremely dangerous and anxiety-provoking act. In the song the women's movements are at a minimum, whereas in the myth they are constantly moving.

> Two *allaparindja* women of the Kamara class originated at Ljalapuntja (Gravel-place). They both had big *tjurunga* on their heads and strings around their necks. One of them was called Allaparindja and the other was called Naierambuma (Red-tail-feather-of-the-cockatoo). From Ljalapuntja they went to Indora (Hot). There they ground seeds. Then they went to Ljipa (String). There they made string. Then they went to Kultjinanga (Arm-string). There they made arm strings. From there they went to Ipija (Rock-hole) and from there to Ndadaquarinja (Round-hill). From there they went to Ikundja Creek and then to Arata (Arata-bush). There they made a long string and then climbed a big hill at Kwalpa (Sand-wallaby). After that they went to Itarka (Itarka-bush) and then to Yuta (Porcupine-grass). There they gathered eagle hawk down and decorated themselves. They had a great deal of string with them. They had many *tjurunga*. They sat on the ground close together. Then they all went right into the earth and became *tjurunga*.

Strehlow[9] has recorded several *alknarintja* myths and songs which are similar to the ones related here.

The Meaning of the Alknarintja *Concept*

It would appear that each totem originally had its own *alknarintja*, that all the women of the totem were the *alknarintja*. First, and most important, the *alknarintja* is a representative of the mother-imago. The "eyes-turn-away-woman" is the mother who resists the demands of her son, and also the person who originally owned the *tjurunga*. The *tjurunga* is a symbolic penis. The mythological *alknarintja* owned the *tjurunga* and then gave them to the men. In the initiation rites, men or women representing *alknarintja* give the symbols of manhood to the initiates.

The Central Australian women sleep on top of their small sons in a position that was described to me as being like that of a man lying on top of a woman during coitus. Anuinga told me several of her dreams. In one of them, two men carrying axes sneaked up behind her. She was so frightened that she awoke, but she fell asleep once more and had the same dream, which again awakened her. Although she was very frightened she did not move, but went to sleep again on top of her small son. Nyiki, her son, was one of the children who dreamt of a female demon with a penis. It is not difficult to recognize Anuinga, his mother, in his dream. The position of the woman on top of the man is greatly feared by the Central Australians. They state that the woman might break the man's penis by sitting on top of it. The old men counsel the young men to make every effort to awaken if they dream of an *alknarintja*, for she might attempt to get on top of them and have intercourse. Were this to happen, either the penis would break, or the young man would die.

The *alknarintja* is the phallic mother—the mother with a penis. The small boys must frequently have erections when their mothers sleep on top of them. However, the penis is attributed to the mother, for has she not taken the masculine position? Not only is the *alknarintja* the phallic mother, she is also the castrator of men. She appears to be avid to regain the penis which she gave up, the *tjurunga* she gave to the men. It is just this object which is used in wooing the *alknarintja* and all other women. The *namatuna*, the small bull-roarer, calls the women to the men. When the sound of the bull-roarer enters into the womb of a woman, she conceives, and so the *alknarintja* is also the pregnant mother. Pregnant women are the only ones allowed to cohabit in the inverted position.

In myths, the *alknarintja* is pictured in two diametrically opposed ways, either sitting absolutely still and refusing to move, or in rapid motion, rushing toward the man who has bewitched her. The immobile *alknarintja* is the result of two constellations of ideas. She is the mother who resists the demands of her son, but she is also the fulfillment of one of the son's wishes because she is resisting the demands of the father by refusing to move. Perhaps the best example of this is the *alknarintja* myth of the Honey Ant totem, in which the *alknarintja* becomes the virgin mother (Song and Myth 1, pp. 125–126). The second picture of the *alknarintja*, in rapid motion, is likewise the result of two constellations of ideas. She is the mother as observed by the son during the primal scene, the rapid motion being that of coitus. This picture of the *alknarintja* also represents a fulfillment of one of the son's wishes. In addition, she is the woman rushing toward him in a highly excited state because he has won her through his magic.

The *alknarintja* is frequently represented as menstruating copiously. It would appear at first that this is another instance of the *alknarintja's* resistance to men, since a menstruating woman must refuse to have intercourse. On the other hand, she is also pictured as having a copious vaginal discharge and as urinating profusely, regarded as signs of great sexual excitement. This seeming contradiction can be resolved by the realization that the *alknarintja* as a menstruating woman is a primary love and anxiety object in Daly's sense.[10]

The bull-roarer is the symbol of the penis once owned by the *alknarintja*. The swinging of the bull-roarer is like the sexual pulsation; indeed, we find the same sort of symbolism in the dreams of American and European analysands. The rapid motion of the *alknarintja* toward the men also represents coitus. The bull-roarer is swung in the performance of the *ilpindja* in the magical belief that like will produce like. This is an instance of magic by analogy.

The natives of the Cape York peninsula have two kinds of bull-roarers. One represents a young girl just prior to the age of puberty, and the other stands for a fully mature young woman. The former corresponds to the *namatuna* (the small bull-roarer), the latter to the *tjurunga mborka* (the large or "body" *tjurunga*) of the Central Australians. They are presented to the boys at different stages of the initiation rite. The mythology associated with these bull-roarers is similar to that of the Central Australians. The bull-roarers are believed to be a husband and wife, and represent the inside of the genitals.[11] The bull-roarer was originally the property of the women because it is symbolic of the girl herself. "The bequeathing of the *moiya* or *pakapaka* [different types of bull-roarer, ed.] by women in the beginning of time, for men to swing, is subtly symbolic of the yielding up by woman of herself to man as also of man's interest in 'what belongs to her'. . . . The swinging of the *moiya* appear to symbolize the awakening of the interest of the girls in a sex-relationship, which is as yet forbidden."[12]

If the above is true, then we can understand why the women are not permitted to see or touch the bull-roarer. Sexual desire, the rhythmic swinging of the bull-roarer, is considered male, and the women are therefore forbidden to see or touch it or the glans penis.

Among the Aranda, an old woman is called a "woman father." From the infantile point of view there is an element of maleness in the concept of mother. Such things as the inverted Oedipus Complex and castration anxiety are universal human responses. The unconscious determinants of the behavior of Central Australian males and their unconscious fantasies, which we have indicated in this chapter as being specifically Central Australian, originate in the particular infantile trauma of the male child sleeping under his mother. Such a custom must go far to strengthen the negative or inverted Oedipal attitude of the adult male, and it is this custom which is specifically at the basis of the mythology of the *alknarintja*.

NOTES

1. B. Spencer and F. J. Gillen, *The Arunta*, 2 vols. (London: Macmillan, 1927), vol. 2, pp. 471–472.
2. Carl Strehlow, *Die Aranda-und Loritjästamme* (Frankfurt: Baer, 1908), vol. 1, pt. 1, p. 97.
3. Free translations of the songs are presented here.
4. In most songs the term "breaking up," when it follows the words "the bull-roarer talks," refers to the bull-roarer itself. There are two possible interpretations for this use of the term. In order to make a bull-roarer, the natives must split a piece of wood. This is the first possible meaning. The words may also mean that, once having been made and used, the bull-roarer splits and is broken. My informants led me to believe that in this song the term should be translated as I have done.
5. The impossibility of this statement was not perceived by the informant.
6. This myth, with some difference in detail, was told to Strehlow, probably by the same informant. Cf. Strehlow, *Die Aranda*, vol. 2, p. 44.
7. No source was given for this song—Editor.
8. A term of opprobrium.
9. Cf. Strehlow, *Die Aranda*, vol. 1, pp. 28–29; vol. 3, pt. I, p. 117.
10. C. D. Daly, "Hindu-Mythologie und Kastrations-Komplex," *Imago* 13, pp. 145–198 (1927).
11. U. H. McConnel, "Myths of the Wikmunkan and Wiknatara Tribes," *Oceania* 6 (1935): 83.
12. Ibid.

6

ILPINDJA

Through an investigation of the *ilpindja*[1] of the Aranda and their neighboring tribes, we may gain some new information about the *alknarintja* and, more specifically, about the erotic strivings of those who attempt to gain the love of women with the attributes of these mythical beings. Although three of the finest field anthropologists were my predecessors in the Aranda area, the word *ilpindja,* as the name of a specific type of love magic, is not mentioned in their publications.

Spencer and Gillen[2] nearly discovered the *ilpindja*. Unfortunately, they did not record any texts. Had they done so, they could not have failed to discover the *ilpindja*, a specific kind of love magic (or poetry) with a mythological background. Strehlow,[3] on the other hand, actually did record one sample of what can formally be regarded as an *ilpindja*. However, he obtained only one variant and did not recognize it as a special type of incantation. Spencer and Gillen speak of certain magical practices performed with the *namatuna* (the small bull-roarer), head bands, and other objects which, according to my informants, are always used in connection with the *ilpindja*.

All the objects used in the performance of the ceremony are first "sung," i.e., certain incantations are sung or recited over them. This endows the objects with magical properties which pass into the one who wears or uses them. I obtained only two of these incantations because I probably did not pay sufficient attention to the subject. The first incantation is sung over the *wallupanpa* (long hair strings) which the men place on the women in order to make them fall in love:

O long string
  Make her love me.
O long string
  She will wear you.
O long string
  Make her love me.

Another song or incantation, which I obtained from Merilkna, is the Fire Stick Song. During the ceremony the fire stick is twirled about in the same way as the small bull-roarer, and for the same purpose: it causes the women to come to the men.

Hold the fire stick.
Hold the boys' bark boomerang.
Hold the fire stick, the water penis, the lightning.
Hold the boys' bark boomerang.
Hold the head band which looks like the rainbow.
Hold the fire stick.
Hold the lightning, the lightning.
Hold the fire stick.
The man stood on the edge of a rock.
He looked into the deep water.
Holding the fire stick, he looked into the water for a long
  time.
He put on the rainbow-colored head band.
Then he saw the hip flesh, the woman.

This incantation is used primarily by the Ilpirra, who tie a string to the fire stick and whirl it about. The sparks fly in all directions. The desired woman sees the man in her dream, and an irresistible force drives her to him.

Most *ilpindja* ceremonies are performed in the following way. A drawing of the desired woman is made in the sand. A pubic tassel is placed on the image. All the objects to be used in the ceremony are placed on the image of the woman, where they are "sung." Having been "sung," they are lifted up and pressed against the stomach of the man performing the *ilpindja*, and the power inherent in the incantations is thus transferred from the objects to the man. The man dons a head band to which is affixed the *namatuna*. He decorates himself with wing feathers and wears a nose bone, paints his body with red ochre, and draws circles with the same substance under his eyes. His forehead is covered with charcoal.

When he is fully attired for the ceremony, he sits down in the center of a circle of men who are helping him. They then "sing" his body. He rises and walks away from them, completely assured that he will be successful. When he walks into the camp, the women see the lightning which surrounds his body and crashes into the ground around him. When the woman for whom he has performed the ceremony sees him, she lies down on her stomach and falls deeply in love.

After he has left the camp, the woman runs into the bush in hopes of meeting him. When she cannot find him she sends a young girl, who acts as a go-between, to search for him. This young girl is called the *amba*. She tells the man that the woman has fallen in love with him. Since it is not considered good manners for the man to admit having performed an *ilpindja*, he says: "How is that? I did not do anything." His answer shows that falling in love is always regarded as the result of the performance of some magical act. The *amba* answers: "Yes you did. The woman saw you approaching, surrounded by lightning, and she fell in love with you." Since it is useless for him to deny his actions any longer, he asks the young girl to arrange a time and place for a meeting with the woman.

The *ilpindja* is usually performed in the foregoing way and is reputed to have the indicated results. However, there are local, totemic, and tribal variations. The Ilpirra, for instance, "sing" a shell which is worn over the penis and is regarded as being magically related to rain and lightning.

The most important *ilpindja* of the Aranda is the one connected with the wanderings of the Tjilpa (Wildcat) ancestors. It is performed at Ltalatuma, the most important totemic center of the Tjilpa. That the most important *ilpindja* should be connected with the Tjilpa is to be expected, since the Tjilpa were the phallic beings of totemic mythology. When a great *ilpindja* is being performed, a ceremonial pole is erected with feathers tied to the top. The men spend many nights sitting around the pole and "singing" it. The pole is believed to change gradually into the image of the man for whom the ceremony is being performed; the pole represents the body, and the feathers, the head. However, it is not the image of the man as he appears in everyday life, but the image of his mystical double.

After the ceremonial pole has been "sung" for several days, the women of the area see a vision or, more often, dream of this "ceremonial pole man." They see him surrounded by a halo of lightning and rush toward him, drawn by the lightning. They are said to "see him with the belly," which means that they see and desire him. A magical shell has been hung on the ceremonial pole. When the man dances, he wears one shell over his penis and another tied to a string around his neck. Lightning springs from these shells. The women follow the lightning from camp to camp. As they draw nearer to the ceremonial grounds, the flashes of lightning become more frequent. When they see lightning all about them, they know that they have reached their destination.

The men "sing" the moon as well as the ceremonial pole and shells. The moon is thereby transformed into a man who drives the women toward the men. The women do not see the moon. They see instead a *leltja* (blood avenger) chasing them, and so they run on toward the men, attracted by the lightning and driven by the moon. Should they try to run in the wrong direction, the moon man drives them back onto the right path.

This *ilpindja* and the results the natives ascribe to it exemplify the libido in its original undifferentiated state. The ceremony attracts *many* women, not a particular woman. Women whom the man is permitted by the class rules to marry, as well as women with whom marriage would be incest, are drawn to him. When one of the forbidden women approaches the man, he asks: "What did you come here for?" The woman must lie in her answer. "I came to see my father," she says, or offers some other excuse. Then the man who performed the *ilpindja* distributes the forbidden women among the men of his group. The women become the legal wives of the men to whom they are given. If the man who performed the ceremony is a strong man, or the Tjilpa chief of Ltalatuma, he may be able to keep three or four women for himself.

The inventor of this *ilpindja* was the great Malpunga himself. He sang it as he wandered to Ltalatuma. As he walked, he subincised himself. *Altala* means "place" and *ltuma* means "cutting," so Ltalatuma is the "Cutting Place." This song was given to me by Papina, a man whose spirit was incarnated at Ltalatuma.

ILPINDJA OF THE TJILPA OF LTALALTUMA

Under the mallee tree the men of the Tjilpa totem sit down and think.

They sit down and think.
One of the men is lame.
He dances about on all fours.
As he dances, he rattles his boomerang and shield.
He dances about on the little plain, rubbing his legs together.
Suddenly he stops! He sees the double rainbow.
He hears the cockatoo talking in the bush.
When the bird moves, the bush moves.
The bush looks double.
The white cockatoo is under a stone.
The men look under the stone.
There is the white cockatoo wearing a neck string.
The men dance around the ant hill wearing their neck strings.
"I am a young man," shouts one of them.
And he dances around the rough hill, the rough hill.
He dances around in a big circle.
On the rock and on the rough hill, he dances around.
There are alknarintja women and women of the Pigeon totem there.
They are frightened by the man and sit very still.
He uses a waddy as a walking stick and walks around them.
He crosses in front of them.
He crosses behind them.
He dances here and there around them.
He holds a ceremonial spear and pushes a dry bush out of his path.
He pushes the porcupine grass out of his way.
He clears his path.
He walks on the hard plain.
His leg makes a shuffling sound as he drags it through the sand.
He goes over the hard ground, dancing along.
He wears the white pubic tassel which makes him flash at the ltala.
The pubic tassel is spread out over his penis.
Gleaming white is his head band.
When he stands up to dance his penis is erect.
The erect penis moves the tassel.
His penis moves as he dances.
He erects a ceremonial pole and hits the ground with it.
He hits the ground with the feathers at the top of the ceremonial pole.

This is how he makes the woman strong.
This is how he makes her an alknarintja who resists the advances of other men.
After he has done this, the woman stays in one place.
She lies on her stomach, her legs stretched out.
She stays in one place resisting the advances of other men.
She stays still and becomes strong in her resolve.
She is now an alknarintja and her body is strong.
She stays in one place leaning on her stick.
She stays under a ti tree bush and leans on her stick.
As she sits down, her vaginal fluid comes out.
As she squats, her vaginal fluid flows out.
He grabs her from behind.
He takes her from behind.
They rub against each other as they wander about.
The alknarintja's stick makes cracking noises as they walk along.
She is menstruating as they walk along and the blood pours out.
She is like half raw meat as they walk along.
The image of the moon is a thing unchanging, permanent.
A phantom comes down from the moon.
He wanders all about and then sits down.
The phantom from the moon sits down.
He sits on the shining red sand hill.
He sits completely immobile.
"I am a young man. I am sitting down," says the phantom.
The phantom dons a neck string.
Two bandicoot tails hang on either side of his beard.
He rubs his body with red ochre.
Fully dressed, he stands erect like a big chief.
He whirls his bull-roarer.
He whirls the small bull-roarer.
The bull-roarer, the girl-catcher, makes whizzing sounds.
The ceremonial pole is lying on the ground.
The top end of the ceremonial pole is on the ground.
The ceremonial pole pierces through the sky.
The small bull-roarer brings all the women to one place like the whirlwind carrying rubbish.
The women are excited.
They are happy as they hurry toward the men.

They skip as they come.
They are so excited that the vaginal fluid flows out of them.
As they skip, the vaginal fluid flows.
There are girls walking about everywhere.
The man who emerged from the ceremonial pole is called "The Red Lizard."
He drives the girl toward the ceremonial pole.
He sits down on the shining red sand hill.
The lightning rises all about him.
The lightning, the water penis, rises [4] all about him.
It rises from the earth.
More and more lightning rises and rises.

The exclamation of the wandering Tjilpa, "I am a young man," a boastful vaunting of male pride (or pride in maleness) is the key phrase of the song. This *ilpindja*, invented by the phallic hero and first sung by him while he was preoccupied with his huge penis, is, in every sense of the word, a phallic song. The words "The penis talks" (literal translation) describe the swaying movements of the erect penis. The origin of the swaying of the *namatuna* (small bull-roarer) and of the words found in so many *ilpindja*, "the bull-roarer talks" and "the bell bird talks," can be sought in these movements.

The lines which refer to the pubic tassel and the head band are placed as a kind of introduction to the line, "The penis talks" (as literally translated). They show the parallelism between the pubic tassel and the head band, and illustrate the trend toward upward displacement. In the line which follows, "He erects a ceremonial pole and hits the ground with it," which describes Malpunga breaking his way past all obstacles with his ceremonial pole (which, as we know from other texts, is really his penis), we have the logical inference from the preceding line.

The description of the male in a state of excitement is interrupted when his partner in this drama of love, the *al-*

knarintja, appears on the scene. Generally, *alknarintja* connotes frigidity, but in the *ilpindja*, *alknarintja* is always connotative of great sexual excitement. Moreover, we are told that the ceremony converts a normal woman into an *alknarintja*, i.e., a woman who behaves like an *alknarintja*, in the usual meaning of the word, toward all men but the one who performed the *ilpindja*.

What we have already discovered in our analysis of dreams and myths is confirmed by these incantations. The concept of the *alknarintja* is ambivalent. Its negative aspect exists because of the repression of the positive attitude. Curiously enough, there are signs of the same ambivalence in the symptoms of sexual excitement described in the song. Although the vaginal secretion, the natural expression of sexual excitement, is described as continually flowing, the *alknarintja* is also menstruating. The two fluids which come from the vagina are regarded as equivalents.

Two old men appear in the song as projections of the singer's personalities. Since the singer is a young man, he must have fathers to aid him in winning the woman and becoming a great chief who can give with princely liberality. The erect stature of the old men and the ceremonial pole which pierces the sky both symbolize erection. The lightning shell is worn to cover the penis. Lightning was mentioned by my informants at other times as a dangerous weapon which crashes down from the sky and is called "the water penis." In the unconscious, the penis appears as a weapon. It is also important to note that the bull-roarer or woman-catcher is usually a magical instrument which produces both thunder and lightning. In this song, thunder and lightning appear as erotic symbols.

The symbolism of the song is in part conscious or allegorical. "Earth" is substituted for "sky." There may well be an unconscious determinant in the choice of the cover word, as in the confusion between "above" and "below." The lightning which actually falls down is said to rise, an expression used more appropriately in reference to the penis.

The supreme importance of the Tjilpa *ilpindja* cannot be denied. However, since my native informants could not agree as to the relative value of the other *ilpindja*, I will make no attempt to classify them in terms of importance. "The Two Women Ilpindja" was sung to me by Wapiti. In the performance of this *ilpindja*, the eagle hawk down, the head band, the *namatuna*, the pubic tassel, the feathers stuck into the hair, and the fur string tied around the arm are all "sung."

THE TWO WOMEN ILPINDJA

The two alknarintja are excited.
They touch their pubic hair.
With their right hands they whirl their bull-roarers.
The women stand in the kuntaji bush and whirl their bull-roarers.
They are standing on a sand hill, a round sand hill.
They put a string in the hole of the namatuna.
They don their head bands.
They grease their head bands.
The namatuna, the head band, breaks.
The women sit down.
Each covers her clitoris with her heels.
They sit on the sand hill, each covering her clitoris.
They sit on the waipi grass and hold their sharp yam sticks.
Their pubic hair and their vaginas are shining.
They have greased them.
The alknarintja looks into her vagina.
"Make it dry. Make it dry!" she exclaims.
"I ought to have a man to live with," she says.
The women are very excited.
They touch their pubic hair with their right hands.
They whirl the namatuna with their right hands.
They stand in the kuntulji bush and whirl the namatuna

They eat the winpiri seed, the winpiri seed.
They crack the seeds with their teeth, with their teeth.
The bell bird with black marks on its breast talks in each clitoris.
The sound of the namatuna has excited them.
They are excited.
Because the bell bird talked, they want to go to the men.
Because of the feeling in their clitores, they want to go to the men.
Because of the feeling in their clitores, they want to go to the men.
The women see an emu.
It has rough soles.
The women sit with their vaginas open under the tjintjila bush.
The women see the yam stick in the men's camp.
There is a fire crackling on a flat rock.
Half the rock is black from the fire.
The women are frightened when they see the fire.
The fear causes them to menstruate.
There is a bush with lumps like those in the flesh of people who commit incest.
The women put green ochre on their noses.
This is the decoration used at the ltala.
A mosquito bites them and they hit it.
Blood comes from the bite.
The flat rock bursts from the heat of the fire.
Chips of stone fly all about.
The smoke rises from the fire.
The women see the burned stone.
"By nightfall we shall reach the big mountain," they say.
Near them, on the mulga bush, are goanas.
On the mulga, on the mulga, the goanas grow.
They hear a bird singing from the mulga bushes as they walk along.
The bird talks.
They are menstruating.
The blood is flowing out.
The blood is running out of them onto the ground.
The earth is soaked with their blood and boggy.
Whenever they lift their feet there is a splashing sound.

The road is so wet that they must cover it with dry brush.
As they walk along, their feet sink into the bog,
Into the bog that they have made with their blood.
They are excited.
They wear their arm strings.
They rub their arm strings with the blood in order to make the men fall in love with them.

This is a description of women in heat. The key to the song is the fact that there is a reference to the women's pubic hair which, of course, means that they are completely genital. However, their genitality is of a male type, as shown by the words, "with the right hand" (i.e., they whirl the *namatuna* with the right hand). The *namatuna* is a symbol of the penis, but in the case of a woman it can be considered a symbol of the clitoris.

The concept of roundness seems to be the unconscious determinant of the introductory lines of the song. There are the round sand hill, the circle in which the *namatuna* is whirled, and finally the circular head band, all of which introduce the shining vagina. At the same time are mentioned a group of noncircular objects, including the waipi grass and the sharp yam stick, which introduce and parallel the clitoris. Since the *namatuna* is in the hands of the women, it is called the *kanta-kanta* (head band).

The tension of the song is increased by repetition of the central theme. It is relieved, to some extent, by such incidents as cracking and eating the seeds. In the line "The bell bird with black marks on its breast talks in each clitoris," the main agent and distinctive feature of all *ilpindja* is introduced. Any song which contains a reference to the bell bird is an *ilpindja*, for the bell bird is the traditional messenger of the man performing the *ilpindja*. In the Tjilpa *ilpindja* its place was taken by the moon man. The bell bird drives the girls toward the man with its booming call. Originally, the bull-roarer may have been called the bell bird, but in any case, the bull-roarer and bell bird have become identified with each other.

The emu with rough soles is probably another phallic symbol. We can assume this because expressions like "rough soles" are often used in curses which nearly always have a sexual significance. Another reason for this assumption is the existence of an almost parallel line, "emu with the crooked penis," in one of the songs of the Emu totem.[5]

The following lines alternate in their unconscious meaning: ". . . the yam stick in the men's camp" and the "emu . . . has rough soles" mean the same thing. They represent the penis, while the tjintjila bush, the open vagina, and the ". . . fire crackling on a flat rock" are symbolic and open representations of the vagina. The crackling fire represents the libido in the urethral or phallic form. An unconscious analogy seems to have been made among the urethral, the seminal, and the vaginal fluids, for the next line states that the women are menstruating and that they are near an itirka bush. Lumps (*itirka*) always refer to incest. The line "Blood comes from the bite" is probably a distorted repetition of the fact that they are menstruating. The bursting rock and the flying chips may symbolize birth. The song concludes with a glorified and magnified description of menstruation. Blood pours out in such quantities that the country is transformed into a swamp. The menstrual blood is also the love potion of the women.

Wapiti carefully explained that this song covered only part of the tradition, and that songs related to other parts of the migration story were used in other kinds of ceremonies. The importance of this *ilpindja* is emphasized by the fact that the man who performs the incantation whirls the large bull-roarer rather than the *namatuna*.

Since it is hardly likely that there should be two unrelated traditions from the same place concerning women, it is probable that this *ilpindja* forms part of the *Alknarintja* Uralpmindji tradition whose song and ceremony have already been described. Strehlow[6] also has recorded a ritual and a song about the two women of Uralpmindji. Although his song contains references to the "bull-roarer talking" and the "menstruating women," it is in all other respects different from the text I have presented here. This illustrates the correctness of Wapiti's remark that many songs are related to one myth.

The following *ilpindja* was given to me by old Yirramba:

EMU OF THE GREAT PLAIN ILPINDJA

Near the alpacha flower stands the emu, the booming breast.
The heart of the man of the Emu totem throbs when he hears the call[7] of the emu.
The heart of the woman of the Emu totem throbs too.
From its nest the clean one stands up with both its legs stiff.
The man of the Emu totem makes a hollow trumpet of the ulumba tree.
With it, he "sings" the woman.
He is a man with incestuous desires.
Beside the road there is a high steep bank.
The emu stands in the shining altjija grass which grows there.
The Adam's apple of the man rattles with his excitement.
He sees the track of the emu.
He recognizes it by the web between the toes.
He sees the long neck of the emu, its Adam's apple moving up and down.
He sees the fat belly and the bones sticking out on either side of it.
The guts of the woman shiver with desire for the man.
She runs to him.
The navel hole, the hole.
O! the navel hole, in the middle of the navel hole.
From the nest, the clean one comes from the nest, from the nest.
The clean one talks as he goes.

The day breaks, the day breaks.
The old emu hears noises.
He looks at the decoration marks of the young emus.
Then the old emu looks at the nest.
A mosquito bites the old emu.
He rises from the nest where he slept with his young ones.
The old emu wakes up.
He makes a booming sound.
The navel strings of the young emus are hanging down.
In the shade of the blood wood tree
In the shade of the blood wood tree
The young emus talk quickly in the egg.
They come out of the good womb.
They are born.
They stretch their necks.
They look about them.
There are marks on the inside of the eggs.
The legs of the young emus are fat on both sides.
The old emu eats charcoal.
His black excrements fall down.
Charcoal falls from the fig tree into the ancestral cave.
Charcoal descends into the cave.
It falls from the fig tree.
And now the rains begin to fall down.
When the old emu was bitten by a mosquito and awakened,
  a pine tree arose where he had slept.
There is a big mulga bush.
The yam speaks.
He speaks too much.
The man of the Emu totem walks along.
There are many prickly bushes sticking together.
They look like one big bush.
He walks with a stoop.
He takes small steps.
He sees many dingos going into the earth at a place where
  there are many small rocks.
The clay under their feet goes, "Crack, crack."
The day is breaking over the sand of the desert.
A dry bush stands up.
It is daybreak and little stones are falling from the rocks.

This is an important and sacred *ilpindja* in which, as in the previous one, the big bull-roarer is used instead of the *namatuna*. The big *tjurunga* is greased for the occasion, a ritual that has special significance in this ceremony, for this *ilpindja* is also called *pamira*, which means "fat belly." Other equivalents for this *ilpindja* are *tjalupalupa* (navel) and *ilpa mara* (good womb). It is evident that the process described in the song is that of birth, and that the objects employed symbolize the uterus.

"The navel hole, the hole. O! the navel hole, in the middle of the navel hole. From the nest, the clean one comes from the nest, from the nest. . . . The young emus talk quickly in the egg. They come out of the good womb." This is a paean of birth. Old Yirramba told me that the line, "good womb talking quickly they come out" (literal translation) refers to the fact that the young birds make a peculiar whistling noise in the egg. For the bird members of the clan, the *ilpintira* (womb) is the egg, while for the human members it is the uterus.

What is the meaning of the song? When taken in conjunction with the "fat belly" aspect, it can only mean one thing—that there is a "multiplication" element in this *ilpindja*. The song may have been sung originally for the purpose of making the emus multiply and grow fat, and then re-elaborated into an *ilpindja*, or it may represent the welding of "multiplication" and *ilpindja* elements. It is not a typical *ilpindja*. There are no references to the bell bird, the moon, or the *namatuna*. Nonetheless, *ilpindja* elements are not completely lacking. The booming noise made by the emu is transformed into the noise made by the hollow trumpet of the man of the Emu totem. The women rush to this man, their guts shivering with excitement. The rattling of the Adam's apple is also a sign of sexual excitement. The long neck is probably a phallic symbol.

It is possible to conjecture at the unconscious motives that led to the synthesis of these two types of incantation. The peculiar habits of the emu must have been observed by these people, for whom it is one of the most important game animals. This bird is remarkable in that it is always the male which incubates the eggs.[8] In the song, the father emu waits for the young birds to emerge from the egg. The man who sings the *ilpindja* is, of course, the emu father calling to himself a great many women who are, in the unconscious, his daughters.

The male who gives birth or tends the eggs (*mbanama* means both "to give birth" and "to lay an egg") is the subject of the song. We cannot pass over the references to the excretory functions. The emu eats charcoal. There is a great interest in the nature of his excrements. That the original meaning of this defecation was birth, and that a displacement subsequently took place, is shown by the fact that the excrements fall into the sacred cave, i.e., into the place from which children come.

I obtained another *ilpindja* of the Emu totem from old Tnyetika, the chief authority among the southern Aranda on tradition and beliefs.

LOVE MAGIC OF THE EMU OF IRPILKA

Irpilka, the home of the man of the Emu totem, is covered with figs.
Irpilka, the home of the man of the Emu totem, is covered with flat rocks.
Many shrubs, all tangled together, cover up the home of the man of the Emu totem.
The flat rock covers up the home of the man.
An inkuta bush also grows upon it.
The young initiates surround it, hiding it from sight.
A tnurunga bush also covers it.
The old man of the Emu totem sees the young initiates walking along.

He puts on his white head ring.
He puts on his white head ring.
He puts on the shining head ring.
The head ring shining blue like the sky.
He puts on the head ring.
The old man of the Emu totem puts on his belt.
He makes ceremonial marks on his forehead.
He ties the bandicoot tails to his belt.
And when the belt is ready, he puts it on.
He covers his body with the shining red ochre.
And when he is all decorated, he stays very still in one place.
His body is shining red and he stays very still, sitting down.
With the red ochre shining on his body, he stays very still sitting down.
With the red ochre shining on his body, he stays very still sitting down.
The indalja feather, which he wears on his back, is moving in the wind.
The indalja feather on his back moves in the wind.
He sees the first little emu.
It has a fat belly.
It is grown up.
He stays at Raknirama sitting very still.
There he sits for a long time, thinking.
He sits on a sand hill and thinks.
He is sitting down on a sand hill and thinking.
The grown-up little emu also sits down.
The women, the women of the fat bellies also sit down.
"Chase them here as we sit and think," he cries.
The women sit down and are very still.
The women are of the wrong class.
O! Lumps in the flesh, you women who sit down!
"Chase them to us as we sit here," the men of the Emu totem cry again.
The old men of the Emu totem puts on his head ring.
He paints round spots on it with the yolk of the emu egg.
He sits down on the nest of leaves that covers the tjurunga.
With his legs stretched out, he sits down on the leaves.
He sits on the flat rock.
On the nest, the covering of the tjurunga, he sits.

He shines like the moon.
He is good like figs!
He is at the sleeping place, the place where the Wildcat ancestors celebrated their inkura.
The huge sand mound extends into the distance.
Dense scrub grows there and extends as far as the eye can see.
He shakes down some seeds from a tree and picks them up.
He picks up the urkurpma seeds, the food of the young emus.
He stands up.
The man of the Emu totem stands on the sand.
He sees an old emu standing up and feeding its young.
The little emus stand in a crack.
Their anuses are full of excrements as they stand in a break in the dense scrub.
The real emu stands up.
Amidst the excrements, he stands up.
He stands up, bending forward like Muti Kutara.
Muti Kutara, the man of the Emu totem, whom the wild dogs chased and who ran away bending forward.
He walks into the dense scrub.
And then he goes right into the rock.
He goes right into a crack in the ground.
Into the rock he goes.

This song contains several lines that are typical of the *ilpindja*. After the old man of the Emu totem sees the country "covered" with initiates, he dresses himself in such a way that he will flash and shine and so attract the women. The arm strings, the belt with the bandicoot tails hanging from it, the shining red ochre, and the feathers swaying in the wind fulfill this purpose. The other lines typical of the *ilpindja* are impossible to mistake. There are many indications that the thoughts of the man of the Emu totem as he sat and thought and cried, "Chase them [the women] here," were about his desires for incest. In this song, as in the first song of the Emu totem, the typical *ilpindja* lines are interrupted by the appearance of a second motif as the old man sees some emus who have just been born. Later in the song the two motifs are combined. The old man decorates his head ring with the yolk of an emu egg. The *tjurunga,* with which he is to perform his magic, are described as being in a nest like the eggs of the emu. And like the male emu, the old man in the song sits on the nest. Just after this he is described as being like the moon, which reminds us of the function of the moon in the Tjilpa *ilpindja.*

The next lines of the song refer to the great *inkura* held by the Tjilpa ancestors.[9] While the man of the Emu totem sings his love incantation, he sees a real emu who is "standing up" and feeding his young. Eggs and birth always indicate the "anal child" fantasy. The next line confirms this. In literal translation it is, ". . . with excrements, stand up stand up. . . ." The song ends with a reference to a myth in which a man of the Emu totem was chased by wild dogs. The ancestor goes right into the earth, down into the sacred cave where he becomes a *tjurunga.*

It is impossible to understand this song fully without knowledge of the myth on which it is based. The first place referred to in the song, Irpilka or Irpulka, is now known as Mount Sunday, and is situated between Henbury and Erldunda. Nearby is a place called Tara (Rock) where there is a small water hole.

> At this place a man of the Emu totem originated. He had two wives, both of whom were *alknarintja* and demons. Many other *alknarintja* women lived nearby. There was another man of the Emu totem living there, who was in love with the wives of the first man. In order to win them, he performed an *ilpindja* and wore arm strings. When one of the wives asked him for whom he had put on arm strings, he replied, "I love you and want you to come with me." He was her *ankalla.* Her husband called her back to him and so she did not go with the other man. (Before that time, the

ancestors of the Emu totem had married their *ankalla*. Since that time, they have not and that is why the custom is still observed in the south.)

When the woman would not go with him, the man of the Emu totem went to the south to gather seeds. He found many and made a big heap of them at a place called Rakneerama (Thinking). He hoped to find someone else of the Emu totem at this place. He actually did find a little emu, who had originated at the spot without egg and without mother.[10] He left the little emu and went back to Irpilka.

The next morning, he again left his camp, but this time he traveled to the east. There he found another small emu at a place called Ltala-lendama (Pulling-lying). There was a feather lying on the ground that had been pulled out of the little emu. Leaving this second little emu behind, he traveled further to the east. At a place called Tneta-ndara (Ground-with-little-holes-in-it) he found a third little emu in one of the little holes. He left the emu there and went back to Irpilka.

The next morning, he again left the camp and went as far east as Puna (Lake). There he went for a swim and then returned to Irpilka. He traveled to the west, searching for seeds, and finally he arrived at a place called Maruqua (Black?). There he met another man of the Emu totem who was his classificatory brother and who was also gathering seeds. The man from Irpilka had found only small seeds, but the other man had found both small and large seeds. He was very cunning, however, and said that he had found only small seeds. He kept the big seeds for himself and hid them in a hole.

The man from Irpilka then ran back to his camp and from there to Maruqua and then back to Irpilka again. Once again he ran to Maruqua and running beyond that place, he came to Alkutanga (From-a-shield). There he found some big seeds and brought them back to his camp. When he got back to his camp, he saw the second man of the Emu totem stealing his big *tjurunga*. He chased this man even as far as Puna, but then he gave up the chase. "Never mind," he said. The second man was an ancestor of the Emu totem who always chased emus with his dogs. The man from Irpilka took the dogs, which the other man had left behind, to Tapara (Back-of-the-emu), a very important place of the Emu totem.[11] Then he went with the dogs to Puna and from there to Irpilka where he had his camp.

A third man of the Emu totem then appeared. He came from Muti-kutara[12] where he was the chief. When he saw the two dogs lying on the ground and showing their teeth, he shouted at them and kicked one of them. Both dogs caught hold of him with their teeth. Their owner arrived at the scene too late to save the third man of the Emu totem. The man ran away with the dogs stuck to his legs. When he grew hungry, he said, "Let me eat some seed." The dogs released his leg and he ran away from them toward the south, but they chased him from both sides. Finally they caught hold of him again, but he ran up a hill to a place called Iliunpa (Emu-smell or perspiration).[13] There he escaped from them by going into the earth and traveling underground. But the dogs waited for him and when he came out of the ground at dawn, they caught him once more. This time he ran westward to Ulpmulpma (Marrow-marrow).

The first man of the Emu totem had been watching the whole struggle. When the third man of the Emu totem saw him, he shouted, "Sit down because of the dogs!" This the first man did. He went right down into the earth and became a *tjurunga*. He is there till this day. Meanwhile, the dogs were hanging onto the other man of the Emu totem. He said to them, "Don't kill me here. It is too far from my home! Rather chase me home and kill me there." [14] But, since the dogs were very hungry and since it was a long way to the man's home, they killed him further to the south at a place called Tnokuta-tnama (Bent-down). They made him bend down and then they killed him.

The dogs ate everything except his bones, his head, and the sacred shell which he wore around his neck. When the dogs had finished eating the meat, they vomited it all up. The meat grew again and went back to the bones. Had the dogs eaten the bones as well, the man would never have been made whole again.[15] After he was whole once more, he stood up and tried to walk about, but he became very giddy and had to sit down again. Finally he found himself a staff and stood up. Even with the staff he could not walk well. He hobbled along for a great distance and finally sat down. A clay pan arose where he sat and is there to this day. He stood up once more and, still leaning on his staff, hopped about like a kangaroo. He became very tired and fell at a place called Ndalkara-lama (Leg-stretch). This was the place where he had originated. He went right into the earth and became a *tjurunga.*

The following song corresponds to a section of the myth which is ethnogically on the borderline between the Aranda and Lurittya traditions.

ILPINDJA FOR IRPILKA AND TARA

He is sitting eternally, but his seat on the ground is uncomfortable.
He is sitting uncomfortably in the corner.
He is sitting off to one side.
He is sitting on their side.
Finally he stands up and moves on.
He holds the back of the yam stick as he climbs.
He climbs higher and higher as he passes on.
He walks, leaning on the back of the yam stick.
Along the path, he moves very quickly.
He moves across the gravel very quickly.
He passes an inkuta bush.
He walks through the tjinta tjinta grass and the inkuta bush.
At the inkuta bush, at that place,
Where the tjinta tjinta grass and the inkuta bush grow, he walks.
He walks past the bush with feathers on his head.
He can smell the cockatoo.
The feathers on his head stand out.
While the smell comes from the creek between the hills,
He walks through the country, making it fertile and making the rains come.
The man of the Emu totem enters a cave when the rains begin to fall.
The cave has many partitions.
He makes a ceremonial mark on his forehead.
He is about to "sing" the women.

He decorates his entire body with red ochre.
He wears a bandicoot tail.
He smears on more red ochre.
He makes marks on his forehead.
He stands on a sand hill, his body smeared with red ochre.
He stands under a fig tree whose boughs spread out over him.
The figs are as big as his belly.
He puts on his head ring.
He ties a bandicoot tail to the end of his ceremonial spear.
He says, "This head ring is mine."
He makes the head ring white and shining.
He does this for the women every day.
He rubs the ring against a tree to make it shine.
He walks along the edge of the cliffs.
He ties his belt tightly so that he will not feel the pangs of hunger.
His arm strings and his belt are shining white.
Because he wears his belt tightly tied, his belly feels full.
With his belt on, he does not feel the pangs of hunger.
His belly feels as though it were full as he walks up and down.
Up and down he walks.
Finally he walks home.
Only the neck of the emu is visible to him.
He hurries back to his camp, dragging the yam stick like a woman.
This ancestor of the Emu totem transforms himself into a robin.
He transforms himself into a red-breasted robin.
He transforms himself into a robin.
He returns to the middle of the place from which he started.
He takes one step and then sits down and does not move.
The place where he sits is soft.
The rubbish flies about as he cleans the place to make a good camp.
He makes a hollow where all the rubbish had been.
He is tired of this camp.
He wants to go to another place where he can make a good camp.
A camp on soft ground with a hollow place.
A camp where all will be soft and the rubbish will fly about.

The myth is constructed like a dream, based on duplication. There are the elements of incestuous love, conflict with the husband, castration anxiety (the loss of the *tjurunga* and the murder by the dogs). Each of these incidents is portrayed by a different actor.

The song itself deals with only one element of the myth. It begins with a description of how the man of the Emu totem sits. He feels uncomfortable and so he rises and, with the aid of a yam stick, walks about upon the hills, coming and going along the same path. As he walks he sees various bushes and smells the scent of the flowers. When it rains, a cave protects him from the downpour.

As he goes along, he makes the country about him "good," bringing rain and making the entire country fertile. This fertility is indicated by the description of figs as large as a man's belly. That he is performing an *ilpindja* is indicated by the fact that he makes marks on his forehead and decorates his body with red ochre. It is characteristic of the *ilpindja* of the Emu totem that the magic radiates beyond human love to all of nature. The love magic of this totem is connected with the fertility of the entire countryside.

Another outstanding feature of the *ilpindja* of the Emu totem is the importance of the theriomorphic elements. Not only does the ancestor walk like an emu, but he actually takes the shape of a robin. The stress on the animal aspect of the totemic ancestor may be related to another feature of this *ilpindja*. The hero of the love song is described as walking like a woman. This is most probably related to the outstanding peculiarity of the emu, the male of which hatches the eggs and cares for the young. The lengthy description of the way in which the man of the Emu totem sits, and the birth elements of the previous song, are the results of the observation of this peculiarity.

On September 8, 1929, I saw the ceremony which corresponds to this myth performed at Middleton Ponds by Tnyetika and Pata-tota of the Aranda-taka, and by Tjina-nyindaka, a Pitjentara. Three figures appeared. The first, portrayed by Tnyetika, represented the man of the Emu totem from Irpilka. He had his entire chest blackened with charcoal. The others had three circles drawn on their chests and two arranged horizontally on their abdomens. The central figure represented an ancestor of the Emu totem, and the other two represented *alknarintja* women who were demons as well. The ancestor attempted to touch the legs of the two *alknarintja*, but they resisted his advances. Finally he managed to put his finger into their "vaginas." He withdrew the finger and smelled it. This was repeated many times. Finally the entire group began to tremble. They were stopped in the usual way, by hugging.

Kanakana, a Mularatara, represented the other man of the story. He pranced onto the scene, trembling throughout his part in the ceremony. He had decorated himself with a series of white marks and circles running down both sides of his body and a white mark on his forehead. This identified him as the wife-stealing man of the Emu totem. In the ceremony he acted out the roles of the other men in the myth, which shows that the three men in the myth represent fragmented aspects of one individual.

The following *ilpindja* was dictated to me by Merilkna, a Ngaratara man, who had many relatives among the more western tribes. Like so many *ilpindja* and *alknarintja* songs and myths, it is associated with Ilpila.

GRASS ILPINDJA OF ILPILA

The bell bird. O! the bell bird.
He makes a booming sound when he talks.
He moves along dragging the two women of the Grass totem with him.

The two women are so excited that their guts tremble.
The birds are talking in the ti tree bush.
As the bell bird and the two women go along they see a kangaroo with white legs.
The kangaroo jumps across a creek.
The kangaroo jumps across the creek to the other side.
The two women tie bandicoot tails to their head bands.
They put on their arm strings.
They rub their head bands on the bark of the gum trees that grow by the side of the creek.
That will make them white.
The bell bird stands next to them.
He holds the spear thrower the way a man does when he goes to court a woman.
He swings the spear thrower.
He is hot.
He splashes himself in the creek.
The women swing the bull-roarer and it talks.
The two women hold long yam sticks.
These are their bull-roarers.
They split the bull-roarer on the outside.
As they swing the bull-roarer, it splits.
The two halves separate.
They find a soft place on the sand.
It is very warm.
Because of the heat, they scrape a hole in the sand.
They sit in the hollow.
On this spot they lie down.
The place is soft there and they sit down suddenly.
They don their tassels.
The tassels hang down about their ears.
Near them are many tjura bushes and a kalala tree.
There are many kalala trees near the big creek.
They sit beside the creek and look behind them.
They make marks on their foreheads with charcoal.
They put on the toadstool.
The head ring is decorated with birds' down.
A streak of black paint runs from their Adam's apples to their navels.
From the Adam's apple, from the Adam's apple runs the streak.

They paint and decorate their head rings until they are very white.
They flash in the sun.
A rat with a tail walks up to them.
His arms are as thin as the arms of a mosquito as he walks up to them.
He wears a pubic tassel, a white one.
The two women see many salt bushes.
And they see the man wearing the pubic tassel, the white one.
The yinpiri bush is covered with flowers.
It looks as though it were covered with birds' down.
But the man, the man is really covered with birds' down.
The young yinpiri bush is covered with seeds.
The man's penis is erect.
He can smell it.
The flies also smell his penis.
The flies smell the odor coming from the erect penis.
They follow him.
A little cloud appears and the rain begins to fall.
The rain is coming down on the ground.
The man stumbles in the dense scrub.
"I can smell the semen on my penis."
"I can smell myself," he says.
The two women are murmuring continually.
They make a noise.
With his right hand, the man catches one of the women.
With his right hand he catches her.
With his hand he catches her.
With his hand he catches her.
She wears a big head ring.
There is a black streak from her Adam's apple to her navel.
His erect penis stretches like a piece of string.
Her urine flows out.
It makes a noise as it flows out and trickles down her leg.
The urine flows out from between her labia.
It makes a noise as it trickles down her leg.
They see a kangaroo and a witjuntu bird.
Leaning on her yam stick, she sits down.
The bell bird talks with his booming voice.

The whistler, the bell bird, has a hard neck.
The bell bird is invisible.
They walk through the grass.
In the grass, their track is invisible.
"Look out behind!"
"There is smoke far off. Look out behind!"
The older woman has a sore under her pubic hair.
She shows her labia to the younger woman, her sister.
The younger sister sees the labia.
Both sisters see the labia of the elder one.
The labia! O, the labia!
The clitoris!
O, the labia, the labia!
There are tassels hanging down around the labia and the clitoris.
There are tassels hanging down.
The long hair strings.
The elder sister carries the long hair string.
She carries her arm string.
She is shivering inside from excitement.
There is a light behind them.
It is the aura of the man who is performing the ilpindja.
He is encircled by lightning.
There is a flash of lightning, a big one.
The women drag their yam sticks behind them.
They drag their yam sticks behind them.
The elder sister says, "I do not want the man.
You can have him.
I will stay where I am."
And so saying, she sits down.
The two women look about them.
They look hard.
They look at each other's genitals.
They look hard at a palm tree.
The elder sister says, "There is a bone in your vagina."
She means that her younger sister is closed, that she is still virgin.
The crow talks, the crow talks.
The crow talks, the crow talks.
The two sisters climb a young spear tree.

They climb up the young spear tree to have a look at the men.
From the top of the tree, they see a cracked spear.
The elder sister has a long split, a long vagina.
The younger sister is a virgin.
The elder sister has a long split, a long vagina.
Their hair is shining.
They see a wallaby who sits up and looks behind him.
It stands up.
The two sisters look at each other and at a malara bush.
The wind blows between their legs.
The elder sister says, "Come along. Hurry!"
They touch the ti tree bush and the malara bush.
They touch the ti tree bush and the elder sister says, "Hurry!"
As they walk, their feet make thudding sounds.
They have gone a long way.
They see the malara bush in the distance.
They pass a young palm tree.
They pass a kutara bush with thorns.
They blow on tinder to make a fire.
They sift their seeds in the wind.
They hear a booming noise.
It is the bell bird, the whistler.
The bell bird talks.

The *ilpindja* does not actually end at this point. The remaining part belonged to Arabi and several other people who lived near Henbury. Nothing I did could induce Merilkna to encroach upon their rights.

The two sisters in this *ilpindja* were women of the Grass totem brought to the men by the bell bird. The song describes the trembling excitement they felt and all the things they saw or heard on their trek toward the men. While the birds sang and a white-legged kangaroo ran across a creek, the women decorated themselves with tassels. The song then returns to the bell bird, who appeared carrying a spear thrower, the symbol of a man seeking a wife. The two women carried yam sticks and a bull-roarer.

Legend has it that originally the *alknarintja* women owned the bull-roarer. One day, when they were whirling it, the string broke. This event appears to have occurred simultaneously at two places: at Uralpminja, which is south of Ilpila, and at Ltalaltuma. Some men camping nearby caught the bull-roarer as it flew through the air. It was actually caught by Malpunga, a man of the Paltara class. The woman who had whirled the bull-roarer was of the Ngaraii class, and therefore stood in the relationship of daughter to him. Malpunga was the great phallic hero whose *tjurunga* was his own penis. We may therefore interpret the statement that the women originally owned the bull-roarer to mean that originally women had penises. In this song and in other *ilpindja* where the women are portrayed with bull-roarers, an equation is made between the male genital in erection and general libidinal excitement.

The noise made by the bull-roarer is paralleled by the sound of the urine trickling down the leg of the woman. The descriptions which follow the relation of this event give a general picture of the women and their surroundings. Some of them, however, seem to have a symbolic significance. Given the phallic meaning of the bull-roarer, the fact that the song is one in which its ownership is attributed to the women, and that in this song the bull-roarer is described as "splitting," we can now make sense of two otherwise meaningless lines. The women climb the yinbara tree, the tree from which spears are made. From the top of the tree they see a spear which had split. This may be interpreted quite simply as, "When the woman mounts the penis, it may split."

From the top of the tree the women looked for the man who had enchanted them. As they looked, each saw the vulva and the labia of the other. In actuality, when two girls become excited, they often manipulate each other's genitals if they cannot find a man to satisfy them. Among the Central Australians, the clitoris and the labia are very important erotogenic zones.

After a description of how the women prepared and decorated themselves for the man, the song relates how the elder sister, a woman of experience, renounced her claims on the man in favor of her younger sister, who was a virgin. The song ends as it began: "The bell bird talks."

The owner of the next *ilpindja* was Wapiti. He stated that the song was also used as a *tjurunga*, i.e., a ritual song.

LOVE MAGIC OF THE RAT WITH THE LONG FOREHEAD

"Put it into the bushes."
"Walk on all fours. Walk on all fours."
Near the creek stands a big ankara tree.
The ankara tree stands beside the creek.
"Look behind it!"
Reflected in the creek he sees many tjuara bushes, many kalala trees and many kalatus.
The creek is very wide.
They stand beside the creek and look behind them.
They make marks with charcoal on their foreheads.
They make marks on their bodies with the black paint of the toadstool.
While they decorate themselves with birds' down and don their head bands, they sing,
"I am a boy with a head band."
They draw black lines on their bodies from the Adam's apple down the abdomen.
From the Adam's apple, from the Adam's apple.
O, from the Adam's apple.
They paint their head bands. They paint their head bands.
They hang rats' tails from their bodies.
They decorate their bodies with the tails of animals.
A man whose arms are as thin as those of the mosquito walks by.
His pubic tassel is white.
As he walks he sees many salt bushes.

He sees many salt bushes.
He walks near the salt bushes, wearing his white pubic tassel.
He passes a young yinpiri bush which looks as though it were covered with birds' down.
The ripe seeds on the bush look like birds' down.
The young yinpiri bush is covered with flowers.
The man makes a spear of a branch of the yinpiri bush.
The spear splits!
A lizard magpie sits on the ground near him.
The man says, "I am always cohabiting."
His spear cracks. It cracks.
What a pity!
What a pity that his spear has broken.
He tries to paint his head band with white paint,
With the white paint of the limewood tree.
He says, "With this white paint, I will try to paint my head band."
With the paint from the limewood tree, he tries to paint his head band.
The smell of perspiration and semen rises from his erect penis.
He smells it.
The flies smell it too.
The flies smell the odor and follow him.
A little cloud appears over his head.
The rain begins to fall.
The rain comes down.
He stumbles and falls on the ground.
He stumbles. He stumbles.
He stumbles continually.
He smells the odor of the semen and the perspiration rising from his penis.
He stumbles and falls because of the dense scrub.
He says, "I can smell myself."
He murmurs to himself as he goes along.
He goes about murmuring continually.
Near him, someone makes a noise.
He sees a woman and catches her with his right hand.
With his hand he catches her.

He wears a big head band.
There is a big mark running from his Adam's apple down his abdomen.
He has an erection.
The public tassel dangles from it.
His penis stretches like a string as his erection gets bigger.
His erect penis stands up like a stick with a knob at the end.
His erection gets bigger and bigger.
There is a wide creek near where he stands.
He walks on a ford across the creek.
He says to the woman, "Vulva, we two will cohabit."
There are red spots on her leg from sleeping too close to the fire.
The flies smell the odor that rises from his penis.
As he cohabits, the flies smell the odor.
They buzz around him, annoying him.
The spear cracks!
The spear cracks!
O, what a pity!
What a pity that his stick has broken.

This is not the complete *ilpindja*, but since the remainder of the song belonged to men of the Paltara and Ngaraii classes of the Aranda near Henbury, my informant would not sing it to me. I therefore obtained the remainder from Old Yirramba.

This first part of the song is not completely comprehensible without reference to the myth to which it is related.

> A man of the Rat totem lived alone at Ullapara. There he built a windbreak. He walked away from his camp and placed a spear in his spear thrower. The spear cracked. As he walked on, he smelled the perspiration on his penis. Finally he came to Ilpila and there, from the top of a hill he saw a woman of the Bird totem. There were several women with her, all of whom were grinding seed. Then all of the women left the camp, leaving only one of their number behind.
>
> The man of the Rat totem, the ancestor, came up behind her and caught her by the wrist. Then he had intercourse

with her. Since the woman was of the Purula class and he of the Ngala, this was an incestuous relationship. Then the man left Ilpila and went on to Tara.

In this *ilpindja,* as in the previous ones, there is the association between "cracking" and "splitting" and the erection of the penis. Incidents which begin with descriptions of sexual excitement and end with the spear or the bull-roarer splitting indicate latent castration anxiety.

The following sections of the song and myth, obtained from Old Yirramba, relate the events which occurred when the ancestor went on to Tara.

LOVE MAGIC OF THE RAT OF TARA

He is decorated with wing feathers.
Good! Good!
He, the father, is decorated.
Good! Good!
The father is excited.
He can feel it in his liver.
He desires the girl.
I, the singer, am sorry for him.
The father is excited and desires the girl,
And I, the singer, am sorry for him.
"In the middle of a place, I catch the girl and pull her along with me."
The father can feel the excitement in his liver,
And I, the singer, am sorry for him.
I am sorry for the father who desires the girl so much.
The girl is always urinating.
The urine is always flowing out of her.
"My song is like a string.
I throw the song and I catch the girl.
My song is like a string
With it I can catch the girl.
I pull the string."
The water is covered with a brackish green scum.
The leaves of the reeds are rustling.
The reeds rustle.

The women see the string in front of their faces.
They are frightened, they flee.
But the man will catch her.
He has the black mark on his forehead.
Far off, the men perform the Itala.
Their feet make thudding sounds as they dance on the hard plain.
The women see a man who wears a red feather.
They fall in love with him.
The feather looks to them like a burning fire.
The man, the father, is decorated with wing feathers.
It is good, good!
The water is salty.

As they walk they pass a round stone.
As they walk they pass a round stone.
They walk to the north.
The girl is with him as they walk to the north.
She cries as she walks along.
The tears fall from her eyes as she walks along.
It is a pity about the spear.
The spear has cracked, it has cracked.
O, what a pity!
Nonetheless the man and the girl walk to the north.
The man, the father says, "Even though the spear has cracked,
I go toward the north."
The flies buzz in front of his face as he cohabits with the girl.
He stops to drive them away.
The girl runs away.
"Stop her!"
His song, the string, flies in front of her face.
"Catch her!"

The girl is sad.
She does not even see her own country.
She sees only the sky.
The man, the father, says, "Nonetheless, I go toward the north."
"Whatever else may happen, I walk toward the north."
He holds her by the arm.

He has captured her, married her.
Together they go toward the north.
He holds her by the arm.
He is so excited that he bites his beard.
His penis becomes like a tree with bark.
They stop to cohabit.
He is so excited that he bites his beard.
He cohabits with the girl who is always urinating,
Whose vaginal fluid is always flowing out.
He cohabits with the girl who is always urinating.
The father ejaculates.
The semen pours out.
The semen keeps flowing out of the father's penis.
His penis pushes past the lips as he ejaculates.
Past the lips, past the mound, his penis pushes in this wild coitus.
"I am ejaculating!" he shouts.

The dancers at the ltala make thudding sounds as they walk on the hard plain.
The hard plain talks with the thuds.

With the girl, the father goes toward the north.
They go a long way.
Her country disappears behind them.
She cries as they walk along.
She sheds tears as they walk along.
She stands in the scrub.
She feels the string of his song blocking her way.
She cannot go back.
Her Adam's apple tingles with excitement and she cannot go back.
The girl is sad and sorrowful to leave her homeland.
But she must follow the man.
Her liver hurts with her sorrow.
But the string of his song forces her to go on.

"With all the magic, we catch the girls.
We catch them," say the men of the Rat totem as they talk among themselves.

The following myth was obtained along with the song from Old Yirramba.

> The man of the Rat totem made his camp at Tara (Rock) near Henbury on the Finke River. Two women camped there, a mother and her daughter. The man made his camp near theirs. All day long he decorated himself with his pubic tassel and his head band. All day long he "sang" the women. One of the women was old and fat. She stood in the relationship of wife to him. The other woman was his classificatory daughter. He "sang" and decorated himself for many days. When the mother and the daughter went to gather seeds, he followed them. When they returned to their camp, he danced for the whole night. The next day, they went to gather seeds once more. This time he caught his classificatory daughter. She cried out, "You are my father! Leave me alone!" But he would not release her. He said only, "Come along. I am going to take you."
>
> He dragged her a little way and then made her lie down. He tried to have intercourse with her, but he failed. He dragged her further away from there and tried again to cohabit with her. Again he failed, for the girl was ashamed. "You are my father," she said. "Leave me alone." But he would not release her. She cried as he dragged her away from the camp.
>
> "Where are you going?" she asked. "North!" The man dragged her on and on against her will. As they went along, he danced, for he was very happy. He had captured a young girl and he was well decorated with bandicoot tails. They went through Nyeingu-kona and finally came to Karilkara (Buck-bush-plain). There they rested. While they were resting, he became so excited that he bit his beard. He rushed at her and they cohabited.
>
> Again he danced as they hurried along. They went through Jay to Iltiraputa (White-stone). As they walked to the north, the girl cried. They climbed a big hill to a place called Ankalla (Cousin) or Itirka or Tjirke-wara (Incestuous). There they camped. The girl cried all the harder as she looked back toward her own country. "We do not have much further to go," said the man. When they came to Worra-tara (Two-boys) he cohabited with her. Then they walked past Tjoritja (Stony-country) to Burt Plain.
>
> Together they made a camp there, although she first tried

to run away. Again he forced her to have intercourse with him. A rock hole called Nura arose on the spot where he put his knee. They then went past Arkaianama (Legs-sit-down, i.e., taboo) and on to Paraltja (Dry-bush). There they sat on the top of a hill and the man showed the girl his own country. "There are many people. Do you see the pine trees? That is where we are going."

They walked a little further and rested at a place called Alakala (Father-and-child). There was another man of the Rat totem at that place. He was performing the last part of the initiation ceremony. One of the initiates saw the man and the girl coming and ran to tell the old man that a man was coming from the south with a captured woman. The old man of the Rat totem decorated himself. He even donned the *wallupanpa* (the long strings of hair).

When the man of the Rat totem who had come from the south with the girl saw the old man and the initiates coming toward them, he knelt down and told the girl to stand on his belt and to hold on to his shoulders. The old man threw his *wallupanpa* around them and dragged them toward him. He placed them in a hole and got in after them. All the initiates followed their mentor, the man, and the girl into the hole. They were all covered with the decorations used in the initiation ceremony. They carried their bull-roarers. They wore their head bands, arm strings, and were covered with birds' down. They all went right into this hole in the earth and became *tjurunga*.

The song and myth just related contain a vivid description of *mbanja* (marriage by capture). Their manifest content is of father-daughter incest. It is significant that this motif appears as a latent element in both the love incantations and the initiation rites.

The following Aranda *ilpindja* is also used as a ceremonial song. It belongs to the tradition of the Boy totem. I obtained it from Old Yirramba.

LOVE MAGIC OF THE BOY TOTEM

The boys are frightened.
A rat has frightened them.

They smell the sweet odor of the hakea.
The boys walk along.
They sit down to rest.
They make headdresses for themselves like those worn by the men.
A big boy comes quite near them.
As they walk along, a crack appears in the ground behind them.
It follows them as they go.
The odor of vaginal secretion rises from the crack in the earth.
The boys, the age mates, smell the odor as it rises from the cleft in the earth.
They turn around.
They return to the north.
They now walk toward the north.
They go toward the north.
Finally they sit down.
They walk one by one.
Like this they walk.
One by one they are subincised.

They meet some girls on the road.
They have intercourse with them.
They fly skyward with the girls.
They fly into the air.
And each has his own girl.
All have girls.
The flies buzz about them.
The flies smell the semen.
The boys smell and lick their own penises.
They are happy as they fly away to the sky.
They strike their waddies together.
It is the fertile season.
The grass, the sex totem of the boys, grows abundantly.
O, boys, boys!
They walk through the dense witchetty scrub.
They push their waddies through the impenetrable scrub.
"This is the creek. This is the creek,"
They repeat as they walk along.

From the blood wood, from the blood wood tree
Which grows near it,
They see the creek.
"Is this the blood wood tree?" they ask.
"Is this the creek?" they ask.
They walk into the dense witchetty scrub.
There is a gap in the bushes.
They strike it with their ceremonial spears.
By the ti tree bushes, they hit the gap.
There a stream arises.
The big boys appear.
They are quite close.
As the boys walk along, the stream which rose from the gap follows them underground.
The boys rise and walk.
They splay their toes.
The boys walk about wearing the headdresses that the men wear.
Finally, they sit down.

For the final rite of the initiation ceremony the neophytes are decorated with charcoal, red ochre, and birds' down. They carry waddies and wear head bands like those worn by the men. They dance in a circle around an old man who carries a ceremonial spear on his back. In this ceremony the blood is taken from the subincision wound and not from the arm.

An *ilpindja* of the Opossum totem was obtained from Merilkna of the Ngaratara tribe. It is used as a love incantation and at the *ltala*.

THE OPOSSUM LOVE INCANTATION[16]

Where the man of the Opossum totem urinated, there is salt on the ground.
An odor rises up.
The penis of the man of the Opossum totem is moving.
His knees, O, his knees!

On his knees he makes a mark with charcoal.
Nearby is a hill of stone.
On the side of this stone stands the white cockatoo bird.
The stone is the mons pubis.
In the dark night, the man of the Opossum totem gathers kempa grass.
He decorates himself with this instead of birds' down.
Decorated with grass, the ancestor walks in a circle.
He sees the ritarita grass growing.
He pulls it up by the roots.
He uses it to decorate himself.
An old man stands in the kempa grass.
The ancestor pulls the grass out by the roots.
The ancestor falls down on the stone.
He falls down in a cave.
He falls down in the grass.
He falls down in a cave.

Urine flows on the ground.
It follows him as he walks to Kanakana.
There are many women at that place.
He is so excited that he urinates.
He sees many bean trees.
The bean trees are bending under the weight of their red beans.
The bean trees are shining red.
Above the pine tree floats a wisp of smoke.
It trembles in the air.
The ancestor, the man of the Opossum totem, must urinate.
He stops walking and does so.
He stops and then he walks on.
The leaves of the gum tree are tied to the legs of the dancers at the Itala.
They make a noise, "Ki ki."
There are seeds on the bean tree.
The wind blows.
The twigs and branches fall off the tree.
They fall on the ground and make a hole.
The man of the Opossum totem cracks the seeds of the bean tree with his teeth.

He cracks the seeds open and swallows the kernels.
He looks at his breast.
On a little plain, on a little plain
Stands an alknalkia tree.
Its flowers tremble when the wind shakes the branches.
Near it is an ant hill.
The man walks and walks.
He passes a big ti tree bush.
There are two ti tree bushes.
They are covered with snow.
He picks the leaves off the bushes and eats them.
He sees a woman.
He lifts up his head band of euro skin and "sings" it.
He decorates his head band with birds' down.
The ancestor masturbates and his penis becomes erect.[17]
His penis stands up.
His anus itches and he defecates.
His penis is erect.
In the grass, in the cave, are the tracks of the opossum.
When he chases the opossum, it says, "Ki ki."

As was mentioned above, this song is used both as a love incantation and at the *ltala*. The *ltala* is the time when wives are ceremonially exchanged and "illicit" love affairs are permitted. Paint marks, which are mentioned in the fifth line of the song (in literal translation), indicate its *ltala* character. While the song is being sung during preparations for the ceremony, cross-cousins decorate each other with paint and "sing" each other's body. This makes the men irresistible to the women, so the song also functions during the *ilpindja*. When the song is sung at times other than at the *ltala*, only the ceremonial head bands are used, and they are not decorated with the *ltala* decorations.

Although the song is handed down as part of the totemic tradition, its use at the *ltala* differentiates it somewhat from other parts of the tradition. However, the song has much in common with the usual songs and myths of the totemic tradition. It details the life and wanderings of the totemic ancestor. While many *ilpindja* are part of a totemic or general tradition, some new ones are added in every generation. These are obtained from the *ngantja*, the double.

I received this *ilpindja* of the Yumu tribe from Urantukutu.

HONEY ANT ILPINDJA

The ancestors of the Honey ant totem walk about on all fours.
They sit down.
They sit down behind the hill.
With their yam sticks, their ceremonial spears,
They break through the sky.
They break through the sky with their yam sticks.
As they go along, a whirlwind passes near them.
The place through which they are going is neither good nor bad.
There they see a woman with a big nose.
The woman is sitting down.
The ancestors, who go about on all fours, sit down near her.
They, the men of the Honey ant totem, sit down.
The initiates wear nose bones.
They dance in a circle around the old man, shouting, "Wa! Wa!"
Near a tnyilingila bush, they clear a space.
They clear a space on an embankment.
They see an alpatakalpa bird.
The work of dancing and clearing the space has made them tired.
The bird has made a hole.
There are many people inside the hole.
There is a great crowd sitting down in this hole.
A great crowd sits down.
The woman with the rough nose, she too sits down.
On the scrub, on the scrub,
He walks on the tips of his toes on the scrub.
The ulpatja bird who walks on the tips of his toes over the scrub growls at the ancestors.

He is angry with them.
The ulpatja bird is growling.
There are many ancestors lying in the hole.
They are preparing to become tjurunga.
They offer each other yalka roots.
Their backs are bent under the weight of the tjurunga which they carry.
They see a woman.[18]
The woman has a rough nose.
She sits down.
It is almost morning.
The day is breaking.
The bird growls at the men of the Honey ant totem.
The people of the place, the men of the Ulpatja totem,
Wear the feathers of the red cockatoo.
The men of the Honey ant totem walk in the direction from which they came.
They see an onion in the ground.
There are many of these onions.
They lie everywhere.
The day has come.[19]
The sun rises.
The mulga, the women's yam stick, splits.
They are at a wet, moist place, a wet, moist place from which. . . .

The following myth is related to this obviously incomplete *ilpindja*.

Many *tala ilpindja atua* (love magic men of the Honey ant totem) went from Patuwaritja (Becoming-a-man) to Putati, and from there to Ajaji (Lime-on-water), going in an easterly direction. They continued in that direction, passing Katapalpa (Big-head) and Wallimpiri (Flat-rock). They then came to Ilili (Rocks) which was located near a creek, and from there went to Unapuna (Dry-rock-hole) and Alkipi (Thud). At Alkipi they heard the thudding noises made by the euros when they run on all fours over the sand. They then went to Walparkiri (Subincision-hole or Penis-hole). There they were frightened by a wind man. This angered them, and they growled at the wind man. When they arrived at Ulumbaura (Grass-on-the-water) they saw grass on the water. When they arrived at Apiljarimi (Ininda-seed) they saw ininda seed on the water. Finally they came to Pupanyi (Lie-on-stomach) and there they went right into the earth and became *tjurunga.*

As is evident from the fact that it does not always tally with the song, the myth is incomplete. It is also difficult to see why this song should be used as an *ilpindja.* We can assume that the complete song and myth contain a description of the intercourse of the men of the Honey ant totem with the woman with the rough nose. The song, therefore, would be used to produce love magic as an "epic incantation," as sympathetic magic. The typical *ilpindja* element, however, the description of the sexual excitement of both of the protagonists, does not appear in this song.

A very typical *ilpindja* and its corresponding myth was obtained from Mulda of the Ngaratara tribe, the Lurittya group that is partially Aranda.

EMU LOVE MAGIC OF THE BLUE PIGEON

They go to the sea.
They go to the shore of the sea.
The man of the Emu totem walks stiffly as they go to the sea shore.
They see a mirage as they hurry along.
They walk stiffly toward the sea.
They step on the salt ground.
Many crickets talk to one another.
The brown cricket, who lives in the porcupine grass, talks.
The thighs of the man and the woman ache.
They are tired.
Again they see the mirage.
They see the mirage.
They walk stiffly.
Their bodies are stiff.
They see the black kupata fruit.
They see the lalitja fruit, the lalitja fruit.

They too are black.
They decorate themselves with the feathers of the emu.
They find the seeds of the pine under the pine trees.
With black paint made of the pine seeds, they make marks upon their ribs.
On their ribs, they paint black marks.
They are in a hurry.
They walk swiftly.
They go toward the sea.
They go quickly.
They see the shore.
They hurry on.
There are stones standing up on the hill.
The woman feels sick inside.
The stone is black, it is black.
The woman feels ill.
She is cohabiting.
She lies on the salt ground and cohabits.
A piece of wood lying on the ground goes into her foot.
The stick pierces her foot.
She stops walking.
The man turns back to look at her.
By the sea, they meet an old man of the Emu totem.
He tells them that the legs of the turkey are bent like a spear thrower.
They see the crooked-legged turkey.
The turkey walks with his legs wide apart.
He walks as though he had a sore between his legs.
The turkey walks down a hill.
It is a real turkey, not an ancestor.
The woman feels sick.
She longs for her home.
She sees the stones standing up on the hill and she feels sick.
The wanderers, the woman and the man, eat figs,
And the woman feels sick with longing for her homeland.
They walk on and on.
The soles of their feet are roughened.
They hear the sound of the sea.
It talks on and on.
The waves of the sea talk.

They walk in a circle, they walk in a circle.
There are rocks where they walk and they walk around them.
There are rocks all about them.
They climb the cliffs by the shore.
They stand on the cliffs.
The cricket who lives on the cliffs talks.
The brown cricket talks.
Near them is a bean tree laden with seed.
The tree falls!
It lies on the ground with its seeds all about it.
The bean seed is scattered everywhere.
The wanderers have found good ground on which to camp.
They shiver with happiness.
The man holds the bull-roarer.
He whirls it about.
A cool wind rises.
The seeds of the bean tree lie on the ground.
The bean tree lies on the ground.
The wanderers walk toward the sea.
The wanderers walk toward the shore of the sea.
The legs of the turkey are bent like a spear thrower.
The turkey stands up.
He is bent like a boomerang.
He stands up, bent like a boomerang, and talks.
It is not a turkey, but a crooked-legged man of the Emu totem.
He is bent like a boomerang and he talks.
He is crooked and bent.
The woman is tired.
She talks.
They are at a place called "In the Testicles."
The woman is tired and she talks.
She too is bent like a boomerang from tiredness.
She is tired inside and she talks.
A cool wind rises.
The grass of the boys' totem grows nearby.
Another kind of grass grows there too.
The beds on which the sick lie, are made from it.
The man talks.
The ground on which they stand is hard.

The woman talks.
They talk on and on.
The distance falls away behind them, as
With short steps they walk along.
They come to Andarandara.
There they find grass.
There they stop.
The man whirls the fire stick with a string.
He uses it like a bull-roarer.
They sit in a circle.
They walk around a blood wood tree.
They wear their head bands as they walk around the blood wood tree.
The sound of the bull-roarer rises to the sky.
It brings the women to the men.
Its roaring is like the sound of thunder.
The sound rises to the sky and brings the women.
It travels a long way through the sky and brings the women to the men.
The bull-roarer draws the women to the men.
There are sparks of fire where the men are.
The men dance round and round and then they sit down.
The light of the fire glows.
The fire shines like the moon.
There are two moon men.
The second arose from the foot prints of the first.
They are the ones who bring the women to the men who perform the ilpindja.
The men sit close together.
There is a rainbow across the sky.
The men wear rainbow decorations across their chests.
The women see the rainbows.
The water penis, the lightning, strikes!
The water penis, the lightning, stands up!
Through the ground, the water penis stands up.
This penis, this erect penis, crashes into the ground.
This penis stands up.
The rim of the penis stands up.
The rim of the penis stands up.
The water penis, the erect penis, the lightning, ejaculates!

It throws the semen about.
The lightning, the water penis, stands up.
It is erect.

As he walks along, his penis hits against the insides of his thighs.
His penis is like the penis, the ceremonial spear, of Malpunga the ancestor of the Wild cat totem.
His penis, the lightning, is erect.
His penis blocks the women with its magic.
It prevents them from running away.
From the sand, from everywhere, the magic arises.
The thunder shakes the ground.
It shakes the women.
It shakes the women to their very eyes.
They look sideways with their eyes.
They gaze about them and they see the father.
Copulate with them!
Eat their vulvas!
Ejaculate into them!
Make them wet!
After cohabiting with them, he hits their legs.
He walks with a springy step.
He is happy.
The wombs of the women shiver with excitement.
Their hearts shiver.
The incantation has done all this to them.
The head band of the man glitters in the eyes of the women.
The head band shines in the eyes of the women.
The women are so excited that they do not know what they are doing.
The cricket talks and their excitement mounts.
They are so excited that they cannot be on their guard.
The men catch them easily.
"There is the cricket who is talking," the women say.
The cricket talks.
He talks loudly.
The bell bird walks on either side of the women.
The incantation floats in front of their faces.
They cannot turn aside from their path toward the men.

A boomerang whizzes by.
It whizzes by and makes a booming noise.
The man who arose from the ceremonial spear sees the thrower of the spear.
It is Nantananta, another man of the Wild cat totem.
The booming of the boomerang rises from the grass, from everywhere.
The women are so excited that they stumble.
The lightning, the water penis, has excited them.
The lightning, O the lightning, stands up.
It crashes from on high.
Round and round whirls the bull-roarer.
The lightning crashes into the earth.
The man is so excited that he bites his beard.
He whirls the bull-roarer and flings it from him.
Near him, is a sulky young girl.
The bull-roarer knocks her down.
It talks, O it talks, as it whizzes through the air.
It lifts her up.

The father sits down.
He is tired.

This long and interesting *ilpindja* appears to have been derived from two different songs. Both a myth of the Wild cat totem and a myth of the Emu totem were given in explanation of it. The Emu version is from the northern Aranda frontier. The Ilpirra, who live still farther to the north, use the song as part of a totemic ritual. In their version, it is the emu who leads the women to the men, a function usually attributed to the bell bird. The emu cohabits with the women before he brings them to the men. The following is the myth of the Emu totem as I obtained it from Mulda.

A man and a woman of the Emu totem went northward from Apilkiri (Blue pigeon). They were newly married. They climbed a sand hill and walked one behind the other at the top. They came to a place where there was a great deal of grass seed. This they gathered and ate. Then they continued walking toward the north. When they came to Imilpi (Quandong-tree) they made two camps and camped separately. In the morning, they found a great deal of lalitja fruit.

The next night, they camped in a claypan on the road and the next day they walked a great distance. Finally, they came to the sea (lake). They crossed the sea and continued northward. When they came to another lake, the woman was tired and sat down. The man continued to walk, but he soon turned around and came back to her. They cohabited. This was the first time that they had had intercourse. The man was very happy.

They went northward again and finally came to a big sand hill. There they had intercourse for the second time. They started out for the north once more, but the woman grew tired and they stopped. Again, they had intercourse. They went on until they came to another big hill and there they had intercourse for the fourth time.

They went beyond the hill and came to a big lake. As they walked through the dense scrub which surrounded the lake, a thorn went into the woman's leg. She became lame and sat down at the next water hole. Finally, they continued on toward the north across the lake. They finally arrived at Kulurpa (Testicles) in Ngali. There, they both went right down into the lake and became *tjurunga.*

The following myth of the Wild cat totem is also used in conjunction with the song just recorded. It was given to me by Mulda, who got it from Yalintjika (He-goes-up), a man of the Wild cat totem who came from Palankinja (Norman's Gulley), a district that is half Aranda and half Kukuta.

At a place called Rangaranta (Dropped), there lived a man of the Wild cat totem whose name was Kilpara (Red-lizard). He may have been the man who arose from the ceremonial spear or he may have been the man who was born when the moon man split. Whoever he was, he later changed his name to Kilpara. He had always lived at Rangaranta.

He performed *ilpindja* continually. Malpunga came from the north with his young men (initiates) and stayed with Kilpara for a time. One day, Malpunga sent the young men out to hunt. One of the youths caught his arm in a crevasse and hung face down on the rock. This is why the place is called Rangaranta (Dropped). Malpunga was watching the young men from Kneritja Tatnama (Father-sits-erect), a round hill in the district.

One day, Malpunga, with his gleaming head band, took all of Kilpara's women away from him and went to the north. Kilpara had all sorts of women: right ones, wrong ones, *alknarintja* women, and women of the Idorida (a little bird) totem. When this happened, Kilpara climbed the hill at Rangaranta. When he got to the top, he made a small ceremonial spear. He donned his head band and made an image of the moon out of euro skin. He put on the long hair strings, and placed tail feathers in his hair. Over all of this, he put a *matati*. He then made *wawilja*. He sat facing the south and "sang" these objects. All the women returned from the north. The specter of the moon acquired a body. He became a man. Then this man split in half and there were two men. The ceremonial spear also became a man. The tail feathers became his head. This man who arose from the ceremonial spear looked and sang just like Kilpara.

The original Kilpara went to Indjarankinja (Many-at-noon). Again he performed an *ilpindja*. A bit to the west, at a place called Alknarintja, there were many *alknarintja* women whom he "sang." They all came to him. He was very happy as he cohabited with them one after the other.

He went back to the man who had arisen from the ceremonial spear. Again he sang and again he went to Indjarankinja and sang there too. A great crowd of women came toward him from Ltalatuma. As he stood on the hill at Rangaranta, he saw one *alknarintja* woman hiding. She was one who had not gone to the north with Malpunga. Kilpara "sang" his own penis. It became long, like a snake. He sent it under the ground and when it came out, it went right into the vagina of the hiding *alknarintja* woman. She sat on it and moved up and down, and so, they had intercourse. When they had finished, his penis returned to Kilpara. He kept this woman as his wife.

> The crowd of women that he had seen in the distance arrived and he had intercourse with every one of them. The man who arose from the ceremonial pole went right into the earth and became a *tjurunga*. Kilpara went to Ilarilara (Close-close) and became a *tjurunga* there. The women all became *tjurunga* at Ltalatuma. Even today, this *ilpindja* often brings the wrong women just as it did in mythological times.

Every *ilpindja* has both a subject, the man who "sings," and an object, the women who is "sung." The characteristic element of the former is duplication of the original. There are usually several representatives of the male libido—the bell bird, the moon man, and the man who arises from the ceremonial spear. The woman who is "sung" is always an *alknarintja*.

The following song is an *ilpindja* of the *alknarintja* women of Ilpila. It too contains references to the bull-roarer and its mythical representative, the bell bird. It is used in the final rite of the *nankuru* ceremony.

> The alknarintja, the alknarintja's body, sits down.
> She stays in one place without moving.
> The short alknarintja sits down.
> She holds a fig tree as she would a ceremonial spear.
> She sees fig trees growing near the road.
> The figs are not yet ripe.
> She says to the other alknarintja woman, "Don't cry.
> Yours will be. You will get the man who is singing the ilpindja for you."
> These alknarintja women live at Ilpila.
> The men who perform the ilpindja for them live at Intankangu.
> They are of the Bell bird totem.
> The other woman sees the man.
> "I don't want anybody. I am going to continue my wanderings," she says.
> The young man sees the alknarintja.
> He runs away.

He is too frightened to speak.
Even in his fright, his penis becomes erect and the semen flows out.
Although he is frightened, he carries the spear thrower.
He is a man in search of women.
The women whirl the bull-roarer and it splits.
They whirl it with one hand.
With only one hand, they whirl it.
And it splits. The bull-roarer splits.
O, it splits.
It talks as they whirl it.
The yam stick, which is their bull-roarer, is long.
And when they whirl it, it talks.
The alknarintja woman withdraws.
She wears a belt around her waist.
She has gathered the boughs of the fig tree.
"Your figs will not be unripe," she says.
"You will get the man who sings for you."
And the other woman responds, "I don't care.
"I want to go on by myself."
They see a tjutupa tree.
They see many tjutupa trees.
The wind has blown a great deal of seed, a huge amount of seed, off the trees.
One of the alknarintja has smoky, blue-black hair.
Her hair is like the feathers of the ulamba lamba bird.
She stands near a linjeriri bush and puts in her nose bone.
All the alknarintja women put in their nose bones.
These women, who wear nose bones, have big round breasts.
Their breasts are fat.
The short one holds the yam stick like this.
They hear the bell bird.
They hear the bell bird.
The bell bird talks.
All the alknarintja women are decorated.
O! They are the alknarintja.
They are working with their fingers.
They are the alknarintja who sit winding string around a small tjurunga.
Near them are the boughs of the fig tree.

They wear belts around their waists.
They sit under a big walee tree.
They are menstruating.
The blood is pouring continually from the mouths of their wombs.
Urine too, pours out of the mouths of their wombs.
One of the alknarintja women stands up to go for food.
Another alknarintja scolds her, "You vagina with sores, you!"
All the alknarintja women scold each other.
The vagina of the one who stood up is closed.
She is a virgin.
Her vagina is like a creek.
Her labia are the shores of the creek.
The closed one, the virgin alknarintja, stands up.
One of the women cries out.
She sees someone creeping up through the milpa bushes.
It must be a demon!
One of the alknarintja women is excited.
A strong odor rises from her vagina.
The others are angry with her.
"Stinking vagina!" they call her.
The bull-roarer talks.
The bull-roarer talks.
There are many bull-roarers and they all talk.
Many bull-roarers, and all talking.
These women have round breasts.
Their breasts are fat inside.
The alknarintja woman sits on the top of a rock.
Her breasts are round.
She has a black skin.
She stays in one place and does not move.
On the top of the rock she sits, and does not move.
Another woman offers her food, which she refuses.
"You, with the excited clitoris!
Why do you leave the food?" the other woman asks.
"You, with the menstruating vagina!
Why do you leave the food?" she demands.
The woman is decorated with bandicoot tails.
They are the widest bandicoot tails.
The woman sees the man.
His penis is erect.

It looks like a flat rock.
His penis stands up.
His penis stands up.
The alknarintja woman is completely decorated.
"You have decorated yourself well," the other women say.
The bandicoot tails hang down from her belt.
An ——[20] also hangs down from her belt.
Her belt goes around her waist.
The alknarintja women walk quickly, they hurry.
The incantation is a string pulling them on.
The woman with the smoky hair makes a ceremonial spear.
She tells the other women to fetch string.
The strings are tied together at the top of the spear.
They are all excited.
Their vaginas begin to stink.
One of the women cries out.
She sees someone creeping through the milpa bushes.
It must be a demon!
The ceremonial spear is a wide one.
The woman, who decorated herself so well, she ties the strings to the top.
The short alknarintja woman holds a yam stick.
She holds the yam stick like this.
They put string on the rock.
As the urine trickles down their flanks, it makes a whistling noise.
The men whirl the bull-roarer on one side.
They whirl the bull-roarer with only one hand.
Alas! The bull-roarer has split!
The short alknarintja woman holds the yam stick.
She holds the yam stick like this.

The following *ilpindja*, the property of the Alknarintja totem, was also obtained from Old Yirramba. It belongs to a mixed Aranda and Lurittya group.

ILPINDJA OF THE ALKNARINTJA OF URALPMINJA

They pass the malee trees that grow beside the road.
They walk along the road and pass the trees.
"You are having intercourse with your sister."

"You are one who commits incest. Incestuous!"
"Sister, you must chase him away."
"Chase him away!"
As they walk along the road, they pass a water hole.
There, the frogs are singing.
They are hitting one another with their waddies.
The frogs, the singers, are moving in the sea.
And the women pass them as they walk along the road.
They hear the singing of the frogs and they perform an ilpindja.
The women sit down.
Each feels the excitement throbbing in her anus.
They stand up and hurry on.
They go so quickly that they push each other.
As they run down a sand hill, they fall.
They make ceremonial marks on their noses with red ochre.
They decorate their feet in the same way.
Their strides lengthen, their tracks advance.
As they walk toward the sea, they see a watjitampa bush arise from the ground.
They walk onward toward the sea, the great salt lake.
Flowers blossom everywhere.
Finally, they see the ceremonial spears of the men of the Opossum totem.
The men dance around the spears which are stuck in the ground.
The men whirl round and round.
The women can hear the sounds of joy made by the people.
The dancers groan with laughter.
The bull-roarer booms like a tempest over the high rocks.
The two women can see the glittering bull-roarer.
Their Adam's apples quiver with excitement.
The women begin their dance.
They shout, "Kekeke!"
The bull-roarer sends a booming note into the air.
It floats down to the women.
As the roaring note reaches them, they see an ulampulampu bird.
The bird follows after them.
Sitting with open wings, the bird watches them.
They walk along the sand hill which rises near the sea.

From the top of the sand hill, they see something rising up from the sea.
It is a dragon!
He rises from the sea.
He uncoils and raises his body.
He walks to the sand hills and sits upon them.
From a great distance, they see the dragon rise from the sea.

This *ilpindja* verifies many of the conclusions we reached in connection with the first *ilpindja* that we discussed. The first conclusion is that coitus with the *alknarintja* women is a "mythical" form of incest. In this song, one of the *alknarintja* women tells her sister to mend her incestuous ways and to chase away her brother, who is also her husband. The concept of incest is made concrete by mention of the itirka bush (in the literal translation). In the first *ilpindja* we interpreted the itirka bush to mean incest.

One of the *alknarintja* in this song commits incest, and the other castigates her for this act. This duality is intrinsic in the concept of the *alknarintja* as the embodiment of desire and repression. In previously related *ilpindja* and *alknarintja* songs, we have seen that the *alknarintja* are conceived of as phallic beings; they are said to have been the original possessors of the bull-roarer. In this song they are represented as performing an *ilpindja*, which, at the present time, is an exclusively male rite. My informants told me that, in mythical times, the *ilpindja* were performed by the women. I was told that all the ceremonial articles and rites, which are today exclusively male possessions and prerogatives, were once owned by the women, particularly by the *alknarintja*.

The description that follows the statement that the women are performing an *ilpindja* shows the real significance of the *ilpindja*. Through the *ilpindja*, the tension of forepleasure finds an outlet in words. The song describes the throbbing that the women feel in the anus. What is actually meant is a throbbing in the vagina. This is an archaic feature of Central Australian language and poetry. Many other instances of this are found: the use of the same word for urine, vaginal secretion, and semen is an example.

In the following verses, well-known symbols of eros appear. Bushes spring up and flowers blossom. At this point in the song the males become the phallic beings, erecting ceremonial spears and dancing around them. The women are rushing toward them. The *ilpindja* is now in the hands of the male. The dragon in the last verse is the real vehicle of male love magic.

These men are members of the Opossum totem. However, we know that the ceremonial spears are used in the ritual of the Tjilpa *ilpindja,* so here we have two separate rituals connected. It should also be noted that the dancing around the ceremonial spear is a rite that properly belongs to the general body of totemic ritual, and not to the *ilpindja.* There are many structural differences between this *ilpindja* and the previous ones. In the other songs there is a tendency toward repetition.[21] This song is a well-balanced and concise poem of a structure that could be termed genital. It contains only the most universal features of the sexual impulse: forepleasure, incest, ambivalence, projection, and childbirth (the blossoming flowers). Moreover, there is a balance between open, unmasked statements and projections. The poem finishes with a statement of the effects of the bull-roarer on the women and a description of the dragon rising from the sea, a most adequate symbolization of an erection.

The following *ilpindja* was owned jointly by Urantukutu and Merilkna and is part of the tradition of the Yumu and Pindupi tribes.

ALKNARINTJA LOVE MAGIC STONE

The kulukulu bird, the blue bird, sits on the yam stick.
The kulukulu bird sits on the yam stick and talks.

The man stands, holding the yam stick in the mbanja position.
The water in the creek splashes.
In the creek the water goes around.
The gum trees growing on the bank of the creek separate the men from the women.
The bull-roarer talks!
The bull-roarer talks!
Alas, the bull-roarer splits!
I myself have split it.
The alknarintja sits on the ground.
She refuses to have anything to do with the men.
The alknarintja sits.
Her hair is a smoky blue color.
She sits on the ground.
She does not move.
She looks only at her pubic mound.
The ground is cold where she sits.
The cold makes her cry and she sniffles as she sits on the ground.
Her vagina is opening.
It is preparing for coitus.
The alknarintja women are masturbating.
They are rubbing their genitals with their heels.
They are shaking themselves.
This is the way that the penis would shake them during coitus.
The women hang themselves from the fig tree.
Their bodies roll against it.
They turn to stone as they hang from the fig tree.

The final lines of this song are not very clear in the word-for-word translation. My informants told me the myth that goes with the song. A précis is included here.

> Two *alknarintja* women lived at Atula (Stone). They went to Tolpa (Sand hill) and from there to Tantjimatatjina (Desert Oak). When the morning came, they whirled their bull-roarers. Then they went to Laritara (Rubbing beans). There they rubbed seed. They went to Kanpurarkna (Fruit-growing-in-grass) and there they turned to stone.

The following song, an *ilpindja* of the Pitjentara tribe, contains many of the typical *ilpindja* motifs. It was obtained from Nyeingu-tjilpa.

Standing on the backbone of the hill, the two women decorate themselves.
As they run along the backbone of the hill, they, the two women, decorate themselves.
They decorate themselves with bandicoot tails.
They decorate themselves with bandicoot tails as they run along.
From the entrance of the womb there comes,
From that entrance there comes out,
Their pubic hair, as black and as long as the tail of a wallaby.
Their pubic hair is as black and as long as the tail of a wallaby.
The older woman pushes her yam stick into the earth.
She drives it into the ground.
She is trying to make a soakage.
With her yam stick, she digs.
It is evening.
The other woman calls, "Elder sister, hurry!"
As they hurry along, the sand talks beneath their feet.
The elder sister drives the yam stick into the earth.
With her yam stick, she digs.
They stand by the creeks and play the oracle game.
As they cross the cliff, they sit down.
A woman comes toward them from the eternal place.
As she walks, she sees a small wallaby sitting up.
Its belly sticks out.
It sits up in a gap between the cliffs.
It stands up.
It bends over, it bends.
The two women rise up.
A shining white kangaroo stands near a beef-wood tree.
It stands near the beef-wood trees.
The shining kangaroo, the white kangaroo, scratches a hole in the earth.

There it will sleep.
Holding the end of her yam stick, the woman sits down.
Near her, is a wituntu bird and a taianpa bush.
She sees the flowers of the taianpa bush.
She walks bent over. Her feet are bent.
She walks thus to hide her tracks.
When the women rise, their vaginas open like creeks.
The younger sister sees the vagina of the elder.
As she stands up, her vagina does not open.
She is a closed one, a virgin.
The elder sister, whose vagina is open like a creek, rises.
The elder sister, whose vagina is open like a creek, rises.
She looks about her for the kangaroo.
She sees the flower of the lunkura bush.
It is white like the tail of the kangaroo.
"O, this is the lunkura bush.
O, this is the flower of the lunkura bush," she cries.
The woman is bound round with ilara grass.
She is bound so that she will stay in one place.
So that she will not go to other men.
She is bound up to her Adam's apple that she may not be excited by other men.
The women see the panga grass.
The panga grass grows all about them.
They see the young eagle hawks flying into the sky.
They are high in the air flying round and round.
A crow rises up from a bush crying, "Aaaaaa!"
The crows shout, "Aaaaaa!"
Someone is coming.
The younger woman smells the vagina of her sister.
The odor is strong.
"I smell it," she says.
"I smell the odor, the stink, of my sister's vagina."
She smells the yam stick and shows it to her sister.
She drives the yam stick into the earth.
She digs in the earth with the yam stick.
They walk past a lime tree.
"Hurry!" they shout to each other.
They are so cold that they cannot hold the end of the yam stick.

This is the myth that corresponds with the preceding song.

Two women of the Root totem and one man went to Mitjuritara (With-a-dam). He was a very thin man of the Kanal (a kind of poisonous insect) totem. Although the women were his wives, he did not cohabit with them. Every day, he and the women hunted for mulga honey. The man always kept the best part of the honey for himself. When they came back to camp after searching for honey, they tied branches to their legs in order to hide their tracks.

They went eastward to Milpa (Bandicoot-tail) and there they decorated themselves with bandicoot tails. Then they went to Ilintana (Ilindja-grass) where they saw ilindja grass. They camped there for a while and then went on to Pujukarata (Smoky-bush) and from there to a creek called Mulataltala (White-nose). There they separated the white, good seed from the bad seed.

At that place, there were many women with yam sticks. One old woman saw the two women coming and walked up to them. They all sat down and talked. The women of that place were grinding yam seed with urine. The wanderers showed them how to do it with water. "Leave the urine for the penis," they said. "Do it with water." Then they left.

They went to Itjipaniki (Narrow-round) which was a narrow gap in the cliffs. There they played the oracle game. They climbed a hill and saw a small wallaby. As they passed through the gap, they saw a kangaroo. Then they passed a spear bush and cut themselves some spears.

They marched along with their yam sticks on their shoulders until they were tired. Then they sat on their yam sticks and rested. Then they climbed another hill and passed through another gap. They walked southward to Wira (Wild-eagle-gully) where they saw many lupukura bushes. Here they met another *alknarintja* woman. They asked her to join them since a man was following them, but she refused. They saw an eagle hawk and many more lunkuru bushes. At sundown, they heard a crow. They camped at Okata (Hole-for-rubbing-seed). There they rubbed seed with their feet.

A man of the Magpie totem lived at Wira. He was constantly trying to cohabit with the *alknarintja* women of that place. He would decorate himself, wear his nose bone, and follow them when they left the camp. He always kept himself well hidden. He would hide on a little hill and wait for the women to separate when they went in search of food. The two women went eastward to Kinka-kutura (Two women) and from there to Maanda (Shut). There the road was blocked and they had to go to the northeast to Pine Gap.

My informants ended the myth at this point. The remainder of the tradition belongs to the Aranda. It was given to me by Old Yirramba as an *ilpindja*, but unfortunately my field notes do not state just which of Old Yirramba's *ilpindja* it is.

I obtained the next song from a member of the Mulatara tribe, Taltalpana (Bending). He was of the Mala totem. He referred to the song as an *ilpindja*, but another of my informants, Pukuti-wara, said that the song could not be used in this way. Since, however, the song has many of the elements of the typical *ilpindja*, I have included it here. The words of the song are a mixture of the Mularatara and Pitjentara dialects. It is called both an *ilpindja* and a *tingari* (ceremonial pole or spear). The latter name refers to the fact that the women in the song own a *tingari*.

THE ILPINDJA OF THE CEREMONIAL SPEAR

The nose bone, put it in!
The nose bone, O the nose bone!
The nose bone, put it in!
They see the kaluta tree.
A splinter goes into the woman's foot.
She stands near the punkati bush wearing a feather in her hair.
The women walk in a line over the hard plain.
They stand under a gum tree shifting about and looking for a cool place to sleep.
They stand under the gum tree carrying their fire sticks.
They break their yam sticks.

They sharpen their broken yam sticks.
As they go along they see an erulanga bush, an erulanga bush.
They see the flowers of the takataka trees.
The women scatter over the countryside.
They see the inarangawuna bird who lives in the mulga bush.
They each pull out one of their front teeth.
They each pull out a front tooth.
The sores on their heads are bleeding from a fight.
"Hurry up!" they call to each other.
"There is water in the clay pan."
They erect a wind break.
They put in their nose bones.
The women see strangers coming from the west.
They gesticulate with their tjurunga in the prescribed manner.
They look about them.
The food is very good.
The women decorate themselves with red paint.
They stand up and walk away.
There is yalka in their bellies.
Yalka is a very nice food.
They make a new camp and lay down to sleep.
They sleep in their new camp.
They lie on their elbows under the ngana bush.
In their new camp, they lie on their elbows.
They drink water.
They urinate into a bag made of skins.
The bull-roarer talks!
The women split a piece of wood and fashion it into a bull-roarer.
They whirl it on a string.
The women quarrel over the food.
One woman hits another on the head.
The blood wells up.
The women are crying, crying.
They scatter the okarita seeds about them.
They tie rat tails to their feet.
A kalukalutuna bush grows on the hard ground.
A warilja bush grows there too.
A spear bush grows nearby and a minyini tree.

They dance around the minyini tree, performing the ceremonies with their bull-roarers.
They rub their menstrual blood on the short yam stick, their bull roarer,
Their ceremonial pole which they made from the spear bush.
They break the ceremonial poles in half with a stone.
They cut the boughs off a branch of the spear bush to make a yam stick.
They crack off the branches.
They wear bandicoot tails hanging down the sides of their faces.
They look into each others' vaginas crying, "Cunnus! Vulva! Labia!"
"We whirl the bull-roarer," they shout.
"We put on pubic coverings," they cry.
The bull-roarer breaks and falls into the grass.
O! Whirl the bull-roarer!
Standing near the minya bush, they smell each others' vaginas.
They walk about leaning on little sticks.
There are clouds overhead as they stand under the mininninamaralu bush.
They perform the ceremonial opening of the veins.
The blood flows out.
O open the veins!
O tie up the arms!
They lie with their heads pressed to the earth under a prickly bush.
They rub the seeds.
Standing near the lalpa bush, they paint the ceremonial poles.
They are like the young men, like the young men.
Cut the vagina!
The women are stiff after the operation.
They divide into several groups.
They have finished the ceremony.
They see a pankuna tree.
The waves splash near the pankuna tree.
The initiates walk about.
The initiates wear their nose bones.
They pass an inkata bush.
They wear bandicoot tails tied to their heads.

They see the kalpi-kalpi fruit growing on the majala bush.
A woman called Leave Me is tired.
She sits under an erultja bush.
She wears a bandicoot tail on her head.
She wears a bandicoot tail on her foot.
With the tails trailing behind her, she walks to the south.
The kangaroo kisses.
The women walk on in a straight row,
As they walk, their thighs rub against their vulvas.
The bandicoot tails hang down.
They whirl the bull-roarer.
They hear the booming.
They throw a boomerang.
The boomerang hits a tree and breaks!
Alas, the boomerang is broken.
The initiates throw the bull-roarer.
They throw it from the top of a hill to the plain below.
They whirl their long hair strings and then throw them too.
They pull their front teeth out and throw them.
With their front teeth out and with their bandicoot tails on their heads, and on their feet,
They walk into a dense scrub.
They see a tjitananka bush, a thick bush.
They see a mauala bush and the kalpi-kalpi fruit on it.
The bull-roarer talks!
The bull-roarer talks!
They see the lumps, the excrements of the iwupa worm.
The iwupa worm is the coat of the ancestors.
The women urinate on the lumps as they walk along.
They walk along, they walk along.
They separate from one another.
They have finished the ceremony.
They go off separately.
They have finished the ceremony.

As Pukuti-wara stated, this song is obviously not an *ilpindja*, but it is given here because it was obtained as an *ilpindja* and because the women in it are obviously *alknarintja*. It is an account of a female initiation ceremony performed by the ancestors. The similarities between this ceremony and the present-day male initiation ceremonies are emphasized, as are the typical *alknarintja* features of women.

Our understanding of the *ilpindja* can best be furthered by a description of the rite in terms of narcissism. In these ceremonies, narcissism acted as the mediator between the ego and the object. A certain amount of healthy narcissism[22] is necessary in the struggle with the environment and in the conquest of the love object. In his preparation for the rite, the objects that he was going to wear became particularly meaningful for the singer. Before donning them he "sang" them, and so endowed them with magic. In our own culture, the care with which the necktie and hat to be worn at the first rendezvous are chosen is similar to this procedure. The Central Australian, however, expressed this psychical process in terms of magic. The *mana* went from him to the object and then back to him when he wore it. This was a purely narcissistic process, if we regard narcissism as that state in which the subject takes himself as a love object.

We find another indication of the narcissism involved in the *ilpindja* in the "splitting." The man who performed the *ilpindja* appeared in at least two, and often more, forms. It was the double, the moon man, bell bird, and so on, who drove the women toward the man. In this way, narcissism mediated between the love object and the ego. These doubles were believed to emanate from the ceremonial pole or the bull-roarer. The bull-roarer was called "penis of the water" (i.e., lightning), or even openly identified with the penis itself. The double who appeared to the women in dreams and visions looked like the man, but had lightning circling around his head. The ceremonial pole, or spear, had the same symbolic significance attributed to spears in all other cultures. Moreover, the greatest of all the *ilpindja* was derived from the ancestors of the Wild cat totem who were phallic beings. Taking all the above into consideration, we are justified in considering the doubles not only as narcissistic representatives of the ego and of libido, but also as personifications of the penis.

The fission of the subject now has a functional meaning. The man who waited for the women to come to him as he sat and "sang" himself and his ornaments represents the infantile phase of development, in which gratification is achieved by "magic," i.e., hallucinatory wish fulfillment. The double, the representative of the penis, who actually drove or led the women toward the man, represents male aggression or active object love. Thus, in the performance of the *ilpindja*, the roles were inverted. The real man of the incantation was passive and infantile, while the fantasy man was active and behaved just as the men did in reality when they were seeking wives.

It is also important to note that the doubles, the links between the ego and the object, in that they were viewed as mythical beings (the moon man, the bell bird), are representatives of the father-imago. The man who performed the incantation identified himself with the fathers who first performed the rites in the mythical period, thus expressing the necessity for an identification with the father in the search for the love object, i.e., the mother.

Having considered the phallic mother aspect of the *alknarintja*, and the fact that the male was in the passive role in the infantile situation (the mother slept on top of her male children), we might expect to find that the males of this culture have passive characters, such as are found in our own culture in those instances where the mothers are particularly masculine. In Central Australia, however, the men have overcompensated for the infantile situation. The fantasy of the phallic mother results in the still more phallic (aggressive) son. In the infantile situation the mother was the seducer; as an adult, the male is an aggressive and even violent seducer, a rapist. We must also consider that the mother evoked erotic sensations in her small son at the same time that she denied his sexual advances. This rejecting aspect was carried over into the adult situation without change. The *alknarintja's* attitude toward the sexual advances of the male was essentially negative.

The three primary features of adult love life among the Central Australians were as follows: (1) object choice was based on the desire for the mother-imago and on a repetition of the infantile situation; (2) in attempting to obtain the love object, the man identified himself with the mythical ancestors, and so with the father-imago; (3) the male was characterized by a sadistic-aggressive, or phallic, attitude.

The sadistic-aggressive attitude of the male is best exemplified by the custom of the *mbanja.* The male attitude is closely paralleled by the masochistic character of the women. Probably the best indication of the women's feelings is that, by applying saliva and charcoal to their arms, they created several patches of extremely soft skin. They did this in order that the men might know that they were expecting to be taken *mbanja,* and also in order to increase the pleasure of the men when they dragged the women into the bush.

Through an understanding of the *ilpindja* and of the *alknarintja* complex, the sexual life and fantasies of the Central Australian peoples have become intelligible. The conflicts common to all men have been dealt with in a manner determined by the specific traumata and infantile experiences of the Central Australians. As the dreams of an individual give us insight into the workings of his psyche, so do the myths and songs of these tribes give us insight into the functions of their culture.

NOTES

1. For previous publications on the *ilpindja,* cf. Géza Róheim, "Women and Their Life in Central Australia," *Journal of the Royal Anthropological Institute* 63 (1933): 226; *The Riddle of the Sphinx* (London:

Hogarth, 1934), pp. 33, 164; Reprint in press. (New York: Harper & Row Torchbooks, 1974). E. H. Davies, "Aboriginal Songs of Central and Southern Australia," *Oceania* 2 (1932): 454–467.

2. B. Spencer and F. J. Gillen, *The Arunta,* 2 vols. (London: Macmillan, 1927).

3. Carl Strehlow, *Die Aranda-und Loritjastämme* (Frankfurt: Baer, 1908), vol. 3, p. 119.

4. In this song "rises" probably means "falls," "from the earth" probably means "from the sky." The charm would lose its efficaciousness if everyone understood it. Therefore, the antonym is substituted.

5. Unpublished myth.

6. Strehlow, *Die Aranda,* vol. 3, pp. 48–52.

7. The emu makes a peculiar noise. "It is a deep hollow sound which nearly all hearers describe in different terms, but which seems to me to be well imitated by drumming on an empty cask with muffled sticks. This peculiar sound is uttered when the cock is coaxing the hen, when the latter is teaching her chicks to feed, and by both birds on most ordinary occasions and especially when there is going to be a change of weather." P. Fountain and T. Ward, *Rambles of an Australian Naturalist* (London: Murray, 1907), p. 146.

8. Fountain and Ward, *Rambles of an Australian Naturalist,* p. 143.

9. It was on that same sand hill where the *inkura* was performed that I learned the song in September 1929.

10. This is a direct translation of the myth as it was told to me. By referring to the text of the preceding song, one can see that the phrase "without egg and without mother" must be an exaggeration of the informant.

11. It is just south of the railway line between Old and New Crown Point.

12. A place west of Henbury.

13. A place west of Charlotte Waters.

14. He did not want to die far from the place where he had originated, lest he should not be able to find his way back to the ancestral cave.

15. J. G. Frazer, *Spirits of the Corn* (London: Macmillan, 1912), p. 256.

16. The setting of this incantation is a place northwest of Ilumbara Maiata.

17. At this point of the ceremony, the performer also masturbates.

18. According to one informant, this is where the *ilpindja* part of the song begins.

19. According to another informant, this is the first line of the *ilpindja* part of the song.

20. This word was not translated in the original manuscript–Editor.

21. Repetition is a very common feature in primitive poetry. See Werner Muensterberger and Hella S. Haasse, *Lyriek der Natuurvolken,* (Arnhem: Van Loghem Slaterus, 1947). (Editor).

22. Paul Federn, "Zur Unterscheidung des gesunden und krankhaften Narzismus," *Imago* 22 (1930): 5.

7

THE SEXUAL LIFE

Before we can discuss the unconscious factors which determine the sexual life of the Central Australians, we must first familiarize ourselves with the actual practices.

Marriage and Jealousy

Although marriage among the Central Australians is polygamic in principle, it is frequently monogamic in practice. Since, as will be discussed, a man is usually betrothed to a girl too young to be married, he will frequently live with another woman until his betrothed grows up. If, at that time, he is desirous of variety in his sexual life, he will bring his betrothed home and find himself the husband of two women. If, however, he finds that he is satisfied with the first woman, or feels that he cannot support two women, he will allow his betrothed to marry some other member of his marriage class and so remain monogamous. He will, even though monogamous, have access to women other than his wife at the *Itala*.

The male has a greater degree of choice in sexual

partners than would appear at first glance. Although the class system makes seven out of eight women taboo to him, in the course of his nomadic wanderings he is sure to find some woman of the right class. It was mentioned in Chapter 1 that he may marry women of at least two of the remaining seven classes with impunity. Temporary and covert sexual relations may be enjoyed with women of at least five of the eight classes without risk of social action.

A Central Australian man has the right to marry any woman of one particular marriage class. Women who stand in this relationship to him are his *noa.* His rights to them extend beyond the mere permission to marry. He has a potential sexual right to each of them, whether or not he or they are already married. His exercise of this right depends largely upon the temperament and fighting abilities of the woman's husband. The husband may allow his wife to go with her seducer, or he may duel with the seducer in order to protect his honor and then allow his wife to leave. If the husband's wish is to keep his wife, he may fight off his rival and then punish his wife to prevent a recurrence of the incident. The husband may also set an ambush and kill the rival outright.

I was told of one incident which illustrates several of the possible modes of action. Aniunga, a young woman, went to fetch water for her husband. A man who stood in the relation of cross-cousin to her was prowling nearby, and tried to rape her. She managed to escape and run back to the camp, but her husband was out hunting. She told Mokutana, a member of her husband's marriage class and therefore her *noa,* of what had happened, since he stood in the relationship of brother to her husband, which made him her natural protector. Mokutana had previously taken a great liking to her and was considering an attempt at exercising his rights as her *noa.* He became jealous when she told him of her experience and refused to believe that she had not had intercourse with her cross-cousin. He cut her with a stone knife as a punishment. When her husband, Tjimbarkna, returned, he believed her story. He tried to spear her cross-cousin and threw a boomerang at Mokutana because he resented his interference.

When a woman attempts to run away with another man, her husband will usually be satisfied with beating her or with cutting her on the leg in order to make her remember her duty to him in the future. I was told, however, of one case in which the punishment was far more drastic. Maliki of the Matuntara group had to wait a long time for the woman whom he had been promised to reach maturity. While waiting, he married another woman. When his betrothed was old enough he brought her to his camp. She and the other woman quarreled and finally the betrothed ran away and went back to her parents. Her father and her brothers were incensed by this and, with the acquiescence of her husband, they killed her.

At the *ltala*, exchange of wives is as much a custom as it is a right. The question of right and wrong is not as important as the question of will a particular husband be too angry, or will the violation of the kinship rules be so extreme and carried out so blatantly that the entire group will be aroused? Another incident which came to my attention will serve to illustrate the attitude of the men at the *ltala*.

Papa-tukutu and Moruntu were brothers. Both were married. At the *ltala*, Papa-tukutu ran off with his brother's wife and kept her as well as his own. No quarrel arose between the brothers. At this *ltala* were two other men who belonged to the same marriage class as the brothers. The two brothers slept with all the women who stood in the relationship of mother-in-law to the four men. Their greediness and selfishness angered the other men and a fight took place. The anger was not aroused by the exercise of the customary *ltala* privilege, but by the particular circumstances.

Many theories have been built upon the assumption that the Central Australian male never becomes jealous. This is patently untrue. Jealousy of the neurotic type, based upon projection of the husband's homosexual desires, is rare and may even be absent. Its absence may be explained by the fact that there are so many other outlets for the homosexual trends. The custom of exchanging wives at the *ltala* has led many authors to believe that the natives do not experience jealousy, but it is probably just this custom, like prostitution in civilized lands, which is an outlet for homosexual desires.

Mbanja *and Betrothal*

It would appear that no clear linguistic distinction is made between marriage and rape. In that many marriages begin with a rape, and even after a marriage has been arranged the only way the man can bring the woman back to his own camp is by force, this linguistic confusion seems to have some basis in reality. The aspect of violence is only rarely absent from the first intercourse of a given couple. The term used to describe this first intercourse or rape is *mbanja*.

Mbanja cannot be translated easily into any European language. The Central Australians use the gesture of a man holding the left wrist of a girl with his right hand to illustrate the term. *Mbanja* means both marriage and intercourse. The relationship between men and women at the *ltala* was described to me as *mbanja*, in which case it meant intercourse and not marriage. A speaker will often add words meaning "they remained married" after saying *mbanja*, in order to avoid ambiguity. One informant described her mother beating another woman with whom her father had had intercourse at the *ltala*, and with whom he had entered into a permanent relationship. She described the relationship between her father and this woman as *mbanja*, meaning both the intercourse at the *ltala* and the marriage which followed it. The meaning of the term is limited, however. It is not used to refer to the intercourse of a husband and wife, nor is it used to refer to the relations which follow a seduction. Generally speaking, the term refers to those instances of coitus in which force was applied by the male.

Uran-tukutu of the Pindupi tribe told me one of his childhood memories. He, his mother, his elder brother, and the girl to whom his elder brother had been betrothed were traveling about together. One day his brother, Puna-tari, tried to catch the girl alone. When she saw him she ran away and held fast to a tree, but Puna-tari beat her arm until she let go. Then he raped her.

Here we see, that even though the girl had been betrothed (*lelindja*) to the man, what passed between them was, in the Central Australian language, *mbanja*, and in English, rape. Before we can understand why the first intercourse of a betrothed couple should require force on the part of the man, we must first understand the betrothal custom itself. When two families wish to be united through marriage, a ceremony is performed between a small girl of one family and a small boy of the other. As a result of this ceremony, the girl becomes the prospective mother-in-law of the boy. It is her duty to provide him with a wife. The man must therefore wait many years for his wife to grow up, since she will be young enough to be his own daughter. Because of his impatience to be married, the man may come for the girl before she or her parents feel that she is ready for marriage.

The use of force can also be interpreted in another

way. As a female child grows up, her future husband's attitude toward her is one of paternal indulgence. The terms used to describe this attitude can best be translated as "petting" or "growing the future wife." There are thus two sides to the relationship between a man and his future wife —indulgence and force, both aspects of the same unconscious complex. The relationship is actually that of a father and daughter, the actual ages of the couple are such as to encourage this concept. The recalcitrance of the girl may be determined by her unconscious fear of incest.

Uran-tukuti spoke of one of his own experiences. When he was at Glen Helen, he saw a young girl gathering grubs. When she saw him she tried to run away, but he was too fast for her. He caught her and raped her, after which she escaped and ran back to her parents. Uran-tukuti stayed in the area and several times was able to surprise the girl in the bush and rape her. Finally she ran away to the Mission, where he could not molest her.

Another informant, Ilpaltalaka, told me of her experiences when she was a young girl and camping with her mother at Yapila. One day, while she was gathering food in the bush, she heard a whistling noise. It was Loatjira, to whom she had been promised, creeping toward her. He wanted her to go away with him, but she ran back to her parents. The next time he appeared she did not run away, but went back to his camp with him.

Tjintjle-wara of the Matuntara tribe stated that when a girl's breasts begin to develop, the men begin to chase her. The girl runs away and hides or stays near her mother for protection. When a man is looking for a woman, he always carries a spear thrower, the meaning of which is simply, "I want you." The woman responds by holding up her yam stick, which means, "But I don't want you." These gestures are not merely symbolic; at times real blows are exchanged.

Folk tales are often as illustrative to the psychologist as narrations of actual experiences. In the following folk tales, we see mirrored the trends and conflicts of everyday life. Mulda of the Merino group told the following story of two *malapakara*. These characters, frequently found in folk tales, are usually inept hunters who are most desirous of women.

> Two *malapakara* were wandering through the country when they felt the presence of a woman. They went a short distance and found a camp. They agreed to kill the man who was staying there and to take the woman for their own. When they saw her going out to gather firewood, they became very excited. Each of them pressed his penis with his hand and said, "We will get her. We will get her tonight." When her husband returned to the camp, he asked them for the name of the place from which they had come. They said that they had come from a distant place. The husband gave them some of his kangaroo meat and the two *malapakara* gave the husband some of their meat. During the night, both men had erections and each told the other to get the woman. The next morning, the husband went to hunt in one direction and the two *malapakara* went in the opposite direction. They walked only until the husband was out of sight and then they returned to the camp. They caught the woman, and, although she resisted and pleaded with them, they took her *mbanja*. Each put his penis into her anus and then over her whole body. The only place that they did not touch was her vagina. When they finished, her whole body was dripping with semen. She finally stood up and wiped off the semen with grass.

In the following folk tale we find another example of *mbanja*.

> Two *malapakara* were wandering together. One of them had a woman with him and camped separately from his friend so that the friend would not see the woman. Nonetheless, the other *malapakara* heard the two of them cohabiting. "What have you got there?" he asked his friend. "I hear you co-

habiting. I hear the noise that your subincised penis makes." "There is no one here," said his friend. "I am only eating meat. That is the noise you hear." The next morning, they went out hunting and the man who had the woman told his friend to go in the opposite direction, but the man was suspicious and sneaked back to the camp. There he found the woman. He dragged her into the bush and cohabited with her from morning to night. When the other *malapakara* returned, he knew that they had been cohabiting because he saw their track. He followed their trail and finally overtook them. Each of the men took hold of one of the woman's arms and, while they pulled her, they hit each other with sticks. The one who had stolen the woman was the stronger man. He killed his antagonist and dragged the woman off into the bush. There he cohabited with her from morning to night.

We must now try to understand why marriage and rape have become almost identical concepts among the Central Australians. In the sexual act there is a biologically conditioned parallelism between the penis and the whole body. The fundamental tendency of the male is to penetrate the female, a goal to be achieved only by the penis. The male has a hallucinatory wish fulfillment through an identification with his own penis.[1] This unconscious identification with the penis is illustrated by the following folk tale.

There was a man, who was a bit like a *malapakara*, and who had teeth on his penis. He would put his penis into rat holes. There the penis would bite the rats and bring them out for the man to eat. As he walked, he sang this song. "Demon snake with your poisonous fangs. Stay away from me." This is the song, sung by most men when they walk, in order to keep away the poisonous snakes. This man sang it in order to keep the poisonous snakes away from his penis when it went under the ground. One day, he put his penis into a hole where there were many rats. The penis caught all the rats and the man tied the rats around his waist, using his penis as a belt. His penis was so long that he could still tie a knot in it after it had gone all the way around his body. Every evening, he ate

> the rats that he had caught. He never used a spear when he went hunting. He used only his penis. One day, the penis went into a hole where there were many snakes. The snakes bit his penis and the poison went right up to his heart and he died.

In the unconscious, the penis is a spear, a weapon. Coitus is therefore rape and the symbol of marriage is a spear. The Aranda call both semen and vaginal secretion *inimba.* A woman has far more *inimba* than a man, for a man has *inimba* only when he cohabits, while a woman has *inimba* all the time. My informants mentioned that when a man spears a woman, she becomes filled with *inimba.* This curious theory can be explained by an unconscious identification of coitus with killing, and of the penis with a spear.

Female Beauty

Before attempting a discussion of coitus itself, we will consider the standards of female beauty held by the Central Australian people. Maliki, an old man and one of the few survivors of the Matuntara group of the Lurittya, described what he regarded as a beautiful woman. She would have fair hair, a "nice" face, big eyes, a pointed nose, round cheeks, round buttocks, round breasts, a fat body, and a fat *mons veneris* with a great deal of hair on it.

The Aranda fashion in female beauty was nearly the same as that described above, save that they liked women whose eyes were small and not too widely opened. Yirramba told me that he wanted a woman with a pretty face, fair hair, black eyebrows, round breasts, big eyes, a small nose, and big buttocks, calves, and *mons veneris.* In general, he said that he liked a woman when there was plenty of her, and became quite excited as he enumerated this list of charms. Lelil-tukutu described female beauty from the Pindupi point of view. In almost all particulars his list was identical with that of Yirramba.

I obtained lists of characteristics considered beautiful in women from several other informants, and in general they corresponded with those already listed. There was throughout an emphasis on the requirement that the woman be fat and large in all aspects, and that there be a good deal of hair on the *mons veneris*.

Coitus

The following lengthy description of a seduction was given to me by Renana and some other Aranda men.

> The man and the woman first looked into each other's eyes. Then the man left the group of people and slyly went off into the bush. The woman saw the man leave, and she too got up and went off into the bush. There they found each other. The woman stood close to the man and they rubbed their breasts together. The man asked the woman if she had come into the bush for his sake and the woman answered that she had come because she loved him. The man then took her into the dense scrub. She opened her legs to him and he put his penis into her vagina. Then the man, who was kneeling, raised the woman, who was lying on the ground, and held her in his arms. The woman held the man with her legs and embraced him with her arms. When they had finished cohabiting, they talked. The man asked the woman if she would come to him again. Then they cohabited once more. They remained with each other. When the man went to another place, the woman went with him. They stayed married to each other. When they returned to their home camp, the people all became very excited. They wanted the couple to separate, but the two loved each other. They were married and would remain so.

We can look at this description from both a sociological and a psychological point of view. In Central Australia, as everywhere else in the world, the eyes pave the way for Cupid, and a look betrays the desires. In this culture, the orthodox and generally recognized form of forepleasure is the rubbing together of the breasts, corresponding to the kiss in our culture. Discussing coitus, Kanakana, a middle-aged man of the Mularatara tribe, explained that his people couldn't cohabit "bending down" (i.e., in the European fashion) because they were afraid that they might place the penis in the rectum. It is easy to infer that he regarded the kneeling position as the natural one, and he stated that, as far as he knew, it was used by all the tribes. Nevertheless, the Central Australians do use other positions. One old woman at the Mission was said to enjoy coitus only when she sat on top of the man, although the use of this position is most vehemently denied and is always associated with anxiety. "The penis might break," was Kanakana's statement. His second explanation as to why this position was not used is also anxiety-laden. He told me that the penis dies during coitus because the vagina is very hot. Other informants mentioned that the inverted position was used at times.

Much stress was placed on the noise made during coitus. It was described as being like the splashing of boggy water. Subincision has the purpose of increasing the size of the penis, the subincised penis looking as though it were double in width. The men attempted to prolong the act and, from reports of the natives at the Mission, some exceptionally virile men appear able to cohabit for as long as half an hour. While we come upon frequent accounts in myths of spontaneous ejaculation, this should not be taken to mean that *ejaculatio praecox* is a common phenomenon. In these accounts the hero was supposed to have been continent for a long time, and under those circumstances the sight of a woman sufficed to cause an ejaculation. Some of my female informants told me that a young man might cohabit as often as three times in one night, but not without an interval of sleep. This was regarded as a maximum.

Coitus *a tergo* was denied, yet it undoubtedly occurred, as shown by the behavior of the children during the play hours and by certain passages in folk tales. Married women were said to take the initiative very frequently. Coitus itself was described as an itching sensation followed by sudden relief. They indicated the act by a rotating motion of the hand.

Sex Totems

Most publications on the Central Australian peoples rightfully emphasize the huge part that totemism plays in their ritual and mundane life. One form of totemism peculiar to the Central Australians is that of sex totems. The sexes are each symbolized by an animal or plant. The girls flirt provocatively by destroying the boys' plant, and the boys do the same with that of the girls.

According to Strehlow,[2] the Lurittya call the boys' plant *mulati* and the girls' plant *okara*. They tease one another by referring to the plants as the twin brother or sister of the other. Among both the Aranda and the Lurittya, a small black bird and a pigeon play the same role. Among the Aranda, a grass called *kwarakaljikaljia* (*kwara* means girl) is the totem of the women. The girls rub the milky sap of this plant on their breasts in order to make them grow quickly. Old Yirramba compared the fluid which flows out of the broken grass to vaginal secretion. The boys' grass, *worakaljia* (*wora* means boy), has a red flower which is supposed to resemble the glans penis. By destroying this plant, the girls are symbolically destroying the penis. Several birds with red feathers or red beaks are also symbolic of the boys' totem. The girls curse the boys by describing the penis in terms of the bird; e.g., "Penis beak of the nyenga bird." A bird with a red breast is the symbol of the girls' totem, and the boys use its characteristics in the same way when cursing the girls.

According to Pukuti-wara and Kanakija, the boys curse the girls by describing their genitals as being like a rock hole. The plant called *kwarakaljikaljia* by the Aranda is called *ngungunguta* (semen-semen or vaginal secretion) by the Lurittya. The name of the boys' totem plant is *alputalputa* (dry-dry). This choice of plant is obviously a result of the Central Australian feeling that females are superior to males, in that their genitals secrete a greater quantity of fluid. Another factor in this choice is that the plant is the totemic representative of the young boys who, of course, have no semen. The boys and girls tease each other in the manner described above. The data I obtained about the Yumu and Pindupi were similar to those already mentioned, except that the girls swear, "Dry totem plant of the penis," and the boys, "Wet totem plant of the cunnus."

These data can be considered in connection with the phallic-sadistic phase of the development of castration anxiety. The boys playfully destroy the plant or bird which symbolizes the female genital, while the girls do the same with the plants representative of the male genital. The sex totem is regarded as the twin or double of all the males or females. This doubling has a libidinal basis and is probably done in an effort to decrease anxiety by denying sexual impulses. We find this mechanism throughout Central Australian culture, e.g., the splitting of the male-imago and the representation of the male ego by the bell bird and the moon man in the *ilpindja*. The Central Australians flirt with each other by pretending not to like the object which represents the genital of the opposite sex. During the play sessions, the boys felt no embarrassment when the other children shouted that they had an erection, but they protested violently when someone suggested that they "put it into the vagina."

Curses

There is a tendency among the Central Australians to deny both their own sexual impulses and the desirability of the genitals of the opposite sex. By far the largest percentage of their curses refer to the genitals; their knowledge of the precise details of anatomy is amazing.

ARANDA CURSES

atna nataka	red and inflamed vulva
atna indita indora	foul-smelling vulva
atna tuta knarra	vulva with a large opening of the womb
atna taraltara	vulva which makes clicking sounds after coitus
atna tjija knarra	vulva with large *labia minora*
atna nira-nira	vulva which always wants coitus
atna arapa knarra	vulva with large *labia majora*
atna ulpura	hollow vulva
atna keltja knarra	vulva with a big slit
atna antaka	wide vulva
alkna tarka tarka	green eyes (those who are shameless enough to cohabit in the daytime see the green bush about them)
atna ndalkauatna	vulva that sees rubbish floating in the creek (those who are shameless enough to go to the creek with the men see rubbish in it when they cohabit
atna ankalankala	vulva which causes lumps in the flesh (incest causes lumps to appear in the flesh
kapita knarra	big head
ula iltolpa	prominent forehead
unta kapita yinaka	your head is caught (i.e., your glans penis is caught in the vagina)

unta ullaka yinaka	your forehead is caught (i.e., the forehead of your penis is caught in the vagina)
atna wondjima	you lick the vulva
injainama atna	you smell the vulva
atna ilkulatana	eat vulva!
atna utra ilkawa	eat excrement!
atna urpula	black vulva
para takia	erect penis
para kurrentora	small penis
para lenja knarra	big glans penis
para kapita knarra	penis with a big head
para tnjana tnjana	penis like a stone for grinding seeds
illita knarra	big testicles
para unja knarra	penis with a big sulcus coronarius
worra kunakuna	bad boy

The following curses are only used in the heat of great anger.

para aralta knarra	penis with a big subincision hole
para injabara	subincised penis
ata ngana katjia ramangaraka	I saw you when you originated (I saw your parents cohabiting)
unda maialila ntajuka	you cohabited with your mother

NGATATARA CURSES

kunna etintina	red vulva
kunna muralla	cohabiting vulva
kunna tukulla	vulva like a rock hole
kunna iltjiriltjira	wrinkled vulva
kunna mangata itikamba	vulva like a half ripe quandong fruit
kunna lapurka	open vulva
kunna noanoa	wide vulva
kalu tukula	penis like a rock hole
kalu muruntu	penis like a water serpent
ngambu puntu	big testicles
kallo kata puntu	penis with a big head
kallo tjiwa tjiwa	penis as big as the flat stone used for grinding seed
kunna tjikintaka nuntu	go and drink from the vulva
tjirara kalikali	crooked leg
ulla kujakuja	bad boy
ngailu nangu nuntu walkantala	I saw you being born (I saw your parents cohabiting)

PINDUPI CURSES

pilji tjitintjitni	red vulva
pilji muralla	cohabiting vulva
pilji tukula	vulva like a rock hole
pilji mangata-mangata	vulva like a ripe quandong fruit (red)
pilji inlapurpa	open vulva
kallo kata talo	penis with a big head
kallo tjiwa tjiwa	penis like a flat rock
tjirara talatala	crooked leg
turi putuputurpa	penis with a rough sulcus coronarius
ngaiulu nangu nuntu mirakatunkunta	I saw you being born
yakunka murula	go and cohabit with your mother!

Since relatively little is known about the curses of the other peoples of the world, it is not possible to make any generalizations about those of the Australians. While curses in most parts of the world are concerned with sex and incest, Central Australian curses are unique in two aspects: the wealth of detail about the anatomy of the genitalia, and the fact that even the desire for coitus is regarded as material for an accusation.

Projection and Denial

Not only do the Central Australian people deny their sexual impulses, they also project them. Perhaps the best examples of this are the remarks made during the ritual group masturbation sessions (to be described below). Pukuti-wara and Kanakana described to me the conversation of cross-cousins. When cross-cousins talk, one will show the other his penis and say: "I have a little one." The other will respond: "Oh no! Yours is as big as that of a demon." Each examines the penis of the other, exclaiming: "Yours is big, mine is little." Other common remarks are: "When you cohabit, yours will be so big that it bursts." "Yours is as hard as a bone." "Yours is like a water serpent." "Your subincision hole is as big as a spear thrower." I was told by another informant that the cross-cousins stand facing each other, each holding his own penis and rubbing it in order to produce an erection. The conversation goes as follows: "You have a big one." "No! Yours is so big that you might kill her completely."

Projection and denial are the mechanisms used most frequently in dealing with sexual impulses. When a group with whom I had not spoken arrived in the district where I was, I asked my informants where I could find them. They replied: "You won't be able to find them. They are probably cohabiting." Whenever I asked my informants whether they were familiar with a particular sexual practice, they denied its existence among members of their own marriage class or tribe, but stated that the members of all other marriage classes and tribes practiced it.

There is one custom of these people that appears related to their excessive use of projection and denial. It frequently happens that a group of men will drag one woman into the bush, where they force her to have sexual intercourse with all of them. Gang rape is known in many parts of the world. In our own culture, the city brothel used to play a parallel role. The aspect peculiar to Australian culture is the attitude of the onlookers, who tease the man who is cohabiting with the woman.

There is a folk tale which illustrates a few of the Central Australian modes of dealing with sexual desires and experiences. This story was told to me by Malkunta of the Ngaratara tribe. A man's wife was stolen by a demon. The man watched the couple cohabiting for a long time before he could muster up the courage to spear the demon. It would appear that the use of projection and denial in these situations (i.e., as defenses against sexual impulses) begins in the infantile situation. These are the defenses employed by the child when he witnesses the primal scene. The emotions and tensions evoked in the child when he watches others having sexual intercourse cannot find adequate gratification or release. Projection is called into play, and the child rids himself of the unwelcome feelings. This is probably the explanation of the otherwise meaningless curse, "I saw you being born," which is obviously a distorted form of "I saw you being made" or "I saw your parents cohabiting."

Voyeurism is frequently a preliminary to coitus. We have already discussed the fact that a group of men will act as an audience while one of their number has intercourse, following which the others will cohabit with the woman. Frequently a man will hide where he can watch two girls masturbate each other. When he has obtained an erection, he will force the girls to have intercourse with him.

Masturbation and Homosexuality

Masturbation and mutual masturbation play an important role in Central Australian culture. Although at first my informants denied all knowledge of the practice among their own group (they admitted that it did occur among the neighboring tribes), with further questioning they revealed a great deal about the practice among themselves. There are three forms which masturbation takes: nonritualized mutual, infantile, and ceremonial or ritualized mutual.

One of the most important forms of nonritualized mutual masturbation is in the relationship of cross-cousins (who, it should be mentioned, are also the principle participants in the ritual mutual masturbation). Some of the masturbatory behavior of cross-cousins has already been discussed in the preceding section. Among the Aranda the brother-in-law is called *antijipana* (grease), because brothers-in-law actually smear one another with grease. This term also means homosexual intercourse. The wife's brother belongs to the same marriage class as the wife, and from that point of view a homosexual relationship with him is admissible. It should also be recalled that in those groups which practice cross-cousin marriage, the brother-in-law is the cross-cousin. Although such homosexual relations are permitted by the marriage class system, an attempt to form such a relationship may meet with a great deal of adverse sentiment. When these customs were discussed, Kanakana said indignantly: "If any man tried to do that to me, I would pull his penis out and then spear him."

Mulda told the story of two men who were inept hunters. They masturbated each other until they both had erections. Then one laughed and said to the other: "You have a big one," to which his friend replied: "No, yours is big." Then they ate and lay down to sleep. One of them got up to urinate, and the second one, seeing him, said: "Oh, my friend, you have a big penis." His penis was like an arm moving up and down. Then the second man said: "We will play with the penis." He walked behind the first man and put his penis into his rectum. The first man looked behind him and said: "What a big penis you have." Then they changed places, and the first man put his penis into the rectum of the second man.

Anal intercourse is not the only form of homosexual relationship. The pubic hair is often used as though it were a vagina. Among the Nambutji tribe, homosexuality is institutionalized. (This word refers to relationships which are socially acceptable and even included in the mores, whereas "ritualized" refers to practices which form a part of the ceremonial life.) After his initiation, every young man becomes the boy-wife of his future father-in-law. This custom exemplifies the general trend of the initiation ceremonies. Libido is displaced from the women to the men, from the mothers to the fathers.

Children often masturbate to ward off anxiety. While watching one initiation ceremony, I noticed that the boy, who was surrounded by a group of men, was playing with his penis. Something was about to happen to his penis, and he was reassuring himself that it was still there and all right; he was consoling himself in an otherwise unpleasant situation. Aldinga, a three year old, was very shy when we first met. He did not want to stand still while I took his picture, which shows him both sucking his thumb and holding his penis. On several occasions I observed his mother playing with his penis.

It was in connection with infantile masturbation that I made inquiries regarding the latency period. While some anthropologists might object to the use I made of the so-called "leading question," I cannot share the opinion that such questions are dangerous. I have never had informants state that customs and institutions existed among them when they did not. When I told my informants that Europeans would behave in a certain way in a given situation and they failed to understand my meaning, I assumed that they had no customs comparable to the one I had mentioned. Often, when my meaning was understood, they would reply: "That may be the way of the white man, but it is not our way."

In just such a way, I told them that European children masturbate when they are quite small, but may give up the practice when they are older. My informants told me that Central Australian children begin to masturbate when they are small and continue to do so throughout their lives. My observations bore out this statement. I cannot be certain that the latency period does not exist at all among them, that there is no decrease and subsequent increase in sexual activity, but I am sure that there is nothing that compares quantitatively to the latency period of European children.

Group masturbation is an essential ritual element, both in mortuary ceremonies and in totemic rites. Before setting out on a blood-avenging expedition, the Pitjentara excite themselves by masturbating one another. The use of group masturbation in the totemic ceremonies is even more striking. In all totemic ceremonies the performers are decorated with birds' down. The down is glued to the body with blood obtained from cuts pricked into the subincision hole with sharp stones or twigs. Since this blood can only be obtained when the penis is erect, group masturbation precedes every totemic ceremony. Unfortunately, we can do little to satisfy our curiosity about the unconscious masturbation fantasies of these people, since such material is almost impossible to obtain.

Oral, Anal, and Phallic Organizations

It will be profitable to examine the sexual practices and ideas from the point of view of the oral, anal, and phallic organizations. Central Australian babies are subjected to a minimum of oral frustration. Traces of this phase survive in the symbolism of dreams, myths, and in mundane life. The literal translation of the question, "*Utna ilkukabaka?*" is, "Have you eaten?" In many circumstances the question means, "Have you had intercourse?" A young girl is described either as *anka* (unripe), unripe for intercourse, or *kalla unma* (ready cooked), grown-up, nubile. *Ngaiala* (hungry) means "hungry for intercourse," and "to drink the vagina" is a euphemism for coitus. In an Ilpirra song there is a line, "Water, drink it!" This refers to the girl with whom the dancer will cohabit after the dance. The "water" is the vaginal discharge.

The oral zone functions during forepleasure, but in the adult it is not an area of independent perversion. The word *aruntjima* can be translated as "kissing," but it is hardly kissing in the European sense of the term. The lips are not compressed. A mother may touch her child with her lips, but she would not kiss it as a European mother would. The word also appears to mean "smelling." The boys described their practice of putting their fingers into a girl's vagina or anus and then smelling them by this term. Touching the vagina with the tongue is also called *aruntjima*. Adults are more or less ashamed of cunnilingus.

Old Renana told me a myth about an ancestor who desired to have intercourse with a woman of the Rat totem. He thrust his stick into a hole in the ground near where she was hiding and then pulled it out. He licked the stick, thinking that it had been in the vagina of the woman. When Renana told this story, the other people present laughed in a knowing way and admitted that cunnilingus was practiced by their people.

Morica, a Nyumu woman, declared that fellatio was a frequent practice. During our play sessions the children were always "smelling" and "kissing" the vagina or the anus of the toy animals. Depitarinja admitted having performed cunnilingus and described it as being "good." I saw a two-year-old boy lift the tail of a dog and kiss its anus.

Oral perversions abound in the life of the child and continue to play a role, though much reduced, in the life of the adult. The child is polymorphously perverse, while the adult is not. In general, I would say that the various perversions function less in forepleasure among the Australians than they do among Europeans. In the adult they certainly do not exist as independent perversions.

It was once believed that coitus *a tergo* was the typical position employed by the Central Australians. This is not the case, although it would be an exaggeration to say that coitus in that position never takes place, or that it is a recent innovation introduced by the Europeans. Coitus in the inverted position is probably more frequent, although both deviations are usually denied. I do not know whether adults ever have anal intercourse in normal heterosexual relationships; however, the practice certainly exists among homosexuals and children. The language testifies to the fact that an unconscious identification is made between the vagina and the anus. In Aranda, *atna,* and in the Lurittya dialects, *kunna,* mean both the vagina and the anus. These terms also refer to excrements.

The phallic phase is probably the most important determinant of Australian sex and fantasy life. In songs, myths, and in general behavior, the odor of urine arouses the sexual impulse. In ceremonies and children's games, urine is the symbol of semen. The characteristic features of the phallic organization were described by Freud.[3] The male child is interested in his own phallus. He attributes a penis not only to all human beings, but to everything alive. It is at this stage that the child becomes frightened that his penis is too small.

Another feature of the phallic phase is sexual curiosity. Through his investigations, the child discovers the vagina and interprets it as a missing penis. It is from this interpretation that the close association between the castration complex and the phallic organization arises. The child is disposed toward homosexuality and tends to deprecate the female genital, seeing the female as a castrated male. There is, however, one exception, one female who is believed to possess a penis: the mother. A being without a penis is considered castrated and punished. The mother, who is both desired and respected, must therefore be endowed with a penis. We have previously interpreted the *alknarintja* as "the mother with a penis." She represents both the desired and the dreaded aspects of this concept.

Besides a tendency to regard women as inferior beings, the phallic phase is also characterized by either an open or a sublimated homosexual tendency. This set of ideas is present throughout Central Australian culture. Not only is there institutionalized homosexuality, but there is also the homosexual character of all rituals. At the same time, women are excluded from all ritual and religious observances.

We can see from the customs practiced on the newborn that the penis and the clitoris are equated. In myths, and in the significant fact that both boys and girls favored phallic objects in their play, we can see that the concept of the female penis is a potent force in Central Australian culture.

We are particularly interested in the penis-weapon equation and in the close connection between the phallic phase and ideas and fears of penetration. The phallic phase is highly narcissistic. The narcissism functions as a protective mechanism against the anxiety evoked by the problem of limiting the amount of sadism and overt aggression carried over from earlier phases of development. The "combined parent" concept is particularly important during this period.[4] This idea that the father's penis is permanently in the mother's vagina is, in part, an attempt to bestow upon the desired and respected mother the penis she lacks.

From the phallic point of view, the penis is a weapon that penetrates into the female. We have seen that "standing with the spear thrower" symbolizes marriage; that marriage is rape; and that the spear is the typical symbol of the penis. Sexual curiosity is closely associated with the phallic organization. It is a form of sublimated penetration into the womb. The religion (or a considerable part of it) of the Central Australian people consists of an attempt to answer the question: "Where do babies come from?"

It is relatively simple to demonstrate that this complex of ideas and phenomena occurs in a given culture area, in other words, to show that the Central Australians, like all other peoples, exhibit in their culture some of the features that characterize the phallic organization. However, it is far more difficult to prove the specific importance of these features in the area in question. In our attempt to do so we shall proceed from the basic description to the libidinal economic problem.

The Castration Complex

My informants denied that children were ever threatened with castration as a punishment for masturbation. As far as I was able to observe, the children frequently masturbated in the presence of adults who raised no objections. However, my informants added that castration was never used as a threat, which appears to be an instance of repression. Pitjentara and Yumu children would threaten each other with such statements as "I will pull your penis out" and "I will cut off your testicles." The children had evidently heard these threats from their parents. By using the threats themselves, they were able to ward off anxiety.

In one folk tale the roles are reversed, and in this way anxiety reduced. The children are the castrators of the giant, i.e., the father.

> There was once a demon who lived with his grandsons. His grandsons played with his testicles, which were as big as pumpkins, lifting them up and then throwing them to the ground. The demon would say, "Be careful, boys. The testicles will fall on you and crush you." When the old man was asleep, the boys pulled at the testicles. They stretched a long way. One of the boys had a stick and hit the testicles with it. The old man awoke and chased the boys, but the boys were able to run faster than the old man and stretched the testicles further and further away from the rest of his body. The old man tried to catch the boys, but they ran between his legs. A few of the boys hung onto the testicles and kept rolling them along. The boys, still holding onto the testicles, ran into a big cave. One of the boys hid at the entrance and when the old man entered the cave, the boy hit him on the forehead and killed him. The boys then burned the old man's body and broke the testicles into many pieces.

We have seen other data to prove that the castration complex exists among the Central Australians. The behavior of the children, their games, analysis of the various *ilpindja*, the role of the sex totems, fantasies and facts about coitus, and the analysis of dreams are among the data at our disposal. The specific importance of the castration complex is shown most fully in the initiation ceremonials. These subincision rites come closer to actual castration than do the usual circumcision rites in many parts of the world. In the case of the Pitjentara medicine men, the rites amount to hemicastration.

There are specific anxieties about coitus which are more prevalent among the Central Australians than among other peoples of the world. The men are afraid that the penis will become stuck in the vagina. They tease each other by saying: "If it takes very long, you will be stuck." They believe that women can hold fast to the penis by hooking the rim of the cervix around the glans penis. It is quite possible that incidents of *penis captivum* are more common with a subincised penis, which is very broad when erect, than with a normal penis. On the other hand, it is tempting to interpret this anxiety as a fear of participation in the primal scene. The validity of this interpretation is attested to by the prominence of *erkurindja* demons in their mythology, and by the connection of the *tjurunga* cult with the primal scene.[5]

Another form of anxiety peculiar to the Central Australians is that connected with the subincision wound. The pubic tassel is worn to hide the subincision hole, not the penis. The men become very embarrassed when a woman sees the wound. If an individual who has not been initiated curses one of the initiated, particularly if the curse has reference to the subincision wound, the punishment is death. During foreplay the women may stroke the penis down toward the glans, but they will never touch the glans itself.

In many ritual songs, the subincision wound is called

"vagina." The blood that flows from it is equated with milk. From this it would appear that the men are hiding their repressed femininity from the women. Because of a return to more infantile ways of thinking, when a woman touches the wound it becomes equated with the vagina. This discovery of their latent passive homosexual tendencies calls forth rage and shame in the men.

The "Big Penis Complex"

Analytic patients often have a great deal of anxiety associated with the belief that their genitals are too small. These feelings originate in the passive-feminine attitude toward the father, and in the castration complex. This same constellation of feelings and ideas is present in the Central Australians, where it is transformed into what might be called the "big penis complex." When European men see one another in the nude, their teasing takes the form of saying: "What a small penis you have." Among the Australians, the accusation is always that the penis is too large. This is another example of the use of projection by these people. The other person is accused of having the sexual desires felt by the accusor. Fables of tribes having very large or very small genitals are related to this complex and to the mechanism of projection. It should be noted that while the anxiety appears to be centered around the fear of having too large a penis, subincision is regarded as a device to increase the size of the penis. Here we have the breakthrough of the original aim.

A story told to me by old Depitarinja, an Aranda from Horseshoe Bend, is illustrative of this complex. The Kaluwiruru, a mythical tribe who were supposed to live far to the west of the Aranda, initiated their young men in the following way. A year after the subincision ceremony had been performed, the second part of the initiation rite took place. A stick was inserted into the anus of the initiate and supposedly drawn out through the penis. Then the leader of the old men told the initiates that they would see many kinds of animals and spear them all, after which they would cohabit with a young girl. When the initiates and the girl returned to the camp, they were to be roasted together in the fire.

This tale was told only to frighten the boys. An old woman told the same story to the girls, assuring them that they would be killed if they cohabited with the boys. However, when the couples actually did return to the camp, their hands were joined by an old man and they were given a fire stick by another old man and an old woman whose bodies had been smeared with red ochre.

What interests us particularly in this story is the fact that the small penis of these people is believed to be due to an operation performed at the initiation ceremony, and that the younger generation is threatened with death as a punishment for performing coitus. In reality, the "operation" performed at birth, in which the penis is pushed back into the foreskin in order that it may not grow too large, is the source of the concept that members of the older generation are castrators.

Universal Phallic Theory, Curiosity about Sex, and the Penis as a Weapon

"Every human being has a penis." According to Freud, this is the content of the phallic organization. "Every human being has a *tjurunga*" is the doctrine of Central Australian theology. The *tjurunga* is a symbolic penis. Among some of our clinical patients, we observe that a reconciliation has taken place between the assumed universality of the penis and the observation that women have no penis. This reconciliation is the theory that women have an invisible penis. The Central Australian woman has a *tjurunga*, just as the men have, only she is not permitted to see it. The boy receives the *tjurunga* after being circumcised as a compensation for what he has lost. Moreover, the *tjurunga* is both a phallic and a "body" symbol; its loss is regarded as a catastrophe.

I hope that in the foregoing pages we have succeeded in proving not only that the phallic organization, as demonstrated in clinical analyses, is found among the Central Australians, but also that it is probably the typical form of sexual organization of that area. In his description of what he calls the "deutero phallic phase of organization," Dr. Jones frequently stresses the anxiety-provoking or neurotic aspects of this phase. He regards it as a neurotic compromise rather than a natural step in sexual development.[6] While there are similarities between the cases he had in mind and our Central Australian findings, there are equally impressive differences. During this period, the sexual curiosity of the boy is turned not to an interest in females, but to comparisons between himself and other males. Dr. Jones, therefore, speaks of a striking absence of an impulse to penetrate, an impulse which would be the result of any curiosity about women.[7] The Central Australians are most different from the phallic cases described by Jones in just this respect. Among the Australians, the penis is a weapon and male aggression abounds.

In the cases considered by Jones, privation on the oral level is regarded as responsible for the difficulty of the task of coping with the parents on the genital level.[8] In Central Australia, this type of privation is at a minimum. In the "neurotic" phallic organization as described by Jones and in what we saw among the Central Australians, there are similar starting points and conflicts, but radically different solutions to the problems. The clinical patients have failed where the children of the desert have succeeded. One reason for this difference lies in the different attitudes of savage and civilized mothers. Loving and yielding mothers have daring sons.

The libidinal organization of the Central Australians may be described as being somewhere between the phallic and the genital. Their object relations are not so much ambivalent as weak. They do not make bad consorts, but they are not romantic lovers. Grief for a dead spouse is not of great duration. These are the only people in the world, as far as I know, who perform love magic, not for the purpose of gaining the love of one woman, but in order to attract women in general. The genital act is exciting, rather than the object, and the phallus or coitus rather than the woman. This is what may be called phallic as opposed to genitality—as we aspire to in Western civilization.

The basis of all object relationships is the Oedipal conflict. We have seen that, for the man, the goal is the mother, the *alknarintja*, who turns her eyes away from the desires of the child. This emphasis on the mother's accessibility is the result of the repression of the infantile libidinal trauma. The mother who was once too close is now too far away, and the rapist was once the little boy lying under his mother. The immobility of the *alknarintja* woman is really the rapid motion of coitus, and the woman who refuses to grant the wishes of her son is the same woman who so readily accepts the advances of the father. We have seen that the only man whom a woman accepts as the father of her children is, in the unconscious, her own father, and that a marriage is happy only if the woman succeeds in identifying her husband with her father.

Sadistic and masochistic tendencies are not absent from the emotional make-up of the Central Australian people, as we have already demonstrated. However, they do not exist as independent perversions. Sadomasochistic perversions are associated with the superego.[9] The main difference between

Central Australians and Europeans lies not in the id, but in the range, depth, and function of the superego. While castration threats evidently are made, they are not provoked by infantile masturbation. Actually, there are no cases of frigidity or of psychosexual impotence.

In summary, we may say that, on the whole, the picture is quite favorable. The sexual life and potency of the Australian male is far more "normal" than the sexuality of the European male. This is undoubtedly due to the fact that he has received fewer of the advantages of education. While he still must repress his incestuous desires, the repression need not be as deep as it is among Europeans, and therefore the sexual impulses are not as fatefully distorted.

NOTES

1. Sandor Ferenczi, *Versuch einer Genitaltheorie* (Leipzig: Internationaler Psychoanalytischer Verlag, 1924), p. 58.
2. Carl Strehlow, *Die Aranda-und Loritjastämme* (Frankfurt: Baer, 1913), vol. 4, pt. 1, p. 98.
3. S. Freud, "The Infantile Genital Organization," *The Standard Edition of the Complete Psychological Works of Sigmund Freud*, ed. and trans. James Strachey, 23 vols. (London: Hogarth, 1953–1964), vol. 19, pp. 141–145.
4. Ernest Jones, "The Phallic Phase," *International Journal of Psycho-Analysis* 14 (1933): 1–33.
5. Cf. Géza Róheim, *The Riddle of the Sphinx* (London: Hogarth, 1934), chap. 1. Reprint in press. (New York: Harper & Row Torchbooks, 1974).
6. Jones, "The Phallic Phase," 15.
7. Ibid., 5.
8. Ibid., 10. Many psychoanalysts would find this point of view far too limited. The variety of conflicts and symptomatology due to early oral deprivation can take many different forms such as addiction, depressive states, anxiety, and perversion. The personality constellation of the Australians does not contradict these observations. (Editor.)
9. F. Alexander, *Psychoanalyse der Gesamtpersönlichkeit* (Leipzig: Internationaler Psychoanalytischer Verlag, 1927).

GLOSSARY

In order to facilitate the reading of this book, a glossary of native Australian terms has been compiled. If not otherwise indicated the terms are of Pitjentara origin.

aldola–west wind
aldoparinja–belonging to the west wind
alknarintja–eyes turn away; the mother with a penis
altala–place
altali (Matuntara)–mother's mother's brother
altjera (Aranda)–ancestor
altjira–mythical being; ancestor
amba–go-between
anka–unripe
ankatja kerintja–taboo talk
antijipana (Aranda)–grease; homosexual intercourse
apili (Matuntara)–son's wife (husband's parents)
arakutya knaripata–woman father
aruntjima–kiss, lick
atna (Aranda)–vagina; anus
banga–the blind one
erintja–devil, demon
ilkuma–eat
ilpa mara–good womb
ilpindja–love magic (or poetry) with a mythological background
ilpintira–womb
inarlinga–spiny ant eater
inimba–semen; vaginal secretion
injainama–smell or lick
inkata–chief
inkata knarra–big chief
inkata kurka–little chief
inkiljl (Matuntara)–maternal or paternal grandmother
iwupa–worm with poisonous sting
jaku (Matuntara)–mother
kalla unma–nubile
katta–son
kamauru–uncle (mother's brother)
kami–mother's mother's brother
kami pakali–paternal or maternal grandmother
kankuru–elder sister
karil–lower part of kangaroo
katu (Matuntara)–father
kerintja–shame, avoidance
kultu–upper part of kangaroo
kunanpiri–bird's excrement
kunarpi–maternal grandfather
kunna (Lurittya)–vagina; anus
kunna nurka tara–vagina with blood
kuntili–aunt (father's sister); mother-in-law (woman speaking)

kuri–wife
kuruna–soul (also demon)
kuta–elder brother
kwara–girl
labarindja–woman who refuses to have anything to do with men
lelindja–betrothed
leltja–blood avenger
ltala–a licentious festival
ltana–ghost
ltuma–cutting
malapakara–inept hunters desirous of women
maku–witchetty grub, larva of Cossus moth
malantu–younger brother
malanytu–younger sister
mama–father
marutu–brother-in-law
mbanja–for marriage
milpa-tunanyi–use the crooked stick
minkai–son's wife (husband's parents)
minma–women
mokunerama (Ngaratara)–to become angry and excited
mokunpa (Ngaratara)–old man; big
nalpi-punganyi–hit leaves
namatuna (Central Aranda)–bull-roarer
nana–erection
nankii–sister's daughter
nanunka–always in search of an erection
ngaiala-hungry
ngantja (Aranda)–double, hidden one
nguntju–mother
noa–potential wife
nyilkna–theft; illicit intercourse
nyiri–smallest variety of yam
okalpi–small wallaby
okari–sister's son
otu tunanyi (Mularatara)–breaking
pamira–fat belly
para takia–penis erect
pitji–bark vessel
tamu–paternal grandfather
tana–wooden vessel used only by women
tarpangu–goes right into
tjalupalupa–navel
tjapa–kind of witchetty grub
tjijanku–little pad
tjirke-wara–lumps of flesh
tjukurpa–ancestor
tjurunga–symbolic emblem used in ceremonies
tmara–bark vessel
tnaputa–Don Juan
toari–sister-in-law
totura–headdress worn by non-Christian natives
tuan–man of importance
tuanyiraka–spiny ant eater
tukutita–mythological ancestors
untala–daughter
wallupanpa–long hair strings
waputu–mother-in-law (wife's parents)
waramba–hut for childbirth
warkuntama–circling totemic ceremonial movement by young men
wati–men
wati tukutita–male ancestors
watjira–cross-cousin
wora–boy
yenkua–old man

INDEX